A WILDER KINGDOM

A WILDER KINGDOM

RETHINKING NATURE IN ZOOS, WILDLIFE PARKS, AND BEYOND

EDITED BY

BEN A. MINTEER AND HARRY W. GREENE

Columbia University Press *New York*

Columbia University Press
Publishers Since 1893
New York   Chichester, West Sussex
cup.columbia.edu

Library of Congress Cataloging-in-Publication Data
Names: Minteer, Ben A., 1969– editor. | Greene, Harry W., 1945– editor.
Title: A wilder kingdom / edited by Ben A. Minteer and Harry W. Greene.
Description: New York : Columbia University Press, [2023] |
Includes bibliographical references and index.
Identifiers: LCCN 2023002714 (print) | LCCN 2023002715 (ebook) |
ISBN 9780231201520 (hardback) | ISBN 9780231201537 (trade paperback) |
ISBN 9780231554145 (ebook)
Subjects: LCSH: Zoos—Management. | Zoo animals—Environmental
enrichment. | Zoo animals—Moral and ethical aspects.
Classification: LCC QL76 .W55 2023 (print) | LCC QL76 (ebook) |
DDC 590.73—dc23/eng/20230208
LC record available at https://lccn.loc.gov/2023002714
LC ebook record available at https://lccn.loc.gov/2023002715

Printed and bound by CPI Group (UK) Ltd, Croydon, CR0 4YY

Cover design: Philip Pascuzzo
Cover photo: Shutterstock

CONTENTS

1 Zoos and the Wild: A Reconsideration 1
BEN A. MINTEER AND HARRY W. GREENE

2 Between Worlds: A Conversation Among the Cranes 31
CURT MEINE

3 Animal Art and the Changing Meanings of the Wild 47
ALISON HAWTHORNE DEMING

4 Can Zoos Connect People with Wildness? 63
SUSAN CLAYTON

5 "Wild" Through an American Indian Historical Analysis 77
KELSEY DAYLE JOHN AND REVA MARIAH SHIELDCHIEF

6 Toward a Wilder Kin-Dom: Why Zoos Must Focus More on Ecological Interactions (with Our Children and Other Biota) Than on Isolated Species 93
GARY PAUL NABHAN

7 This Is a Zoo? Reflections on a Wilder Zoo by Visitors to the Arizona-Sonora Desert Museum 107
DEBRA COLODNER, CRAIG IVANYI, AND CASSANDRA LYON

8 Evolution to the Rescue: Natural Selection Can Help Captive Populations Adapt to a Changing World 125

JONATHAN B. LOSOS

9 Zoo Dogs 139

CLIVE D. L. WYNNE AND HOLLY G. MOLINARO

10 Zoo Time 155

NIGEL ROTHFELS

11 The Microbial Zoo: How Small Is Wild? 169

IRUS BRAVERMAN

12 A Home for the Wild: Architecture in the Zoo 183

NATASCHA MEUSER

13 Reconnecting Zoos to the Wild and Rethinking Dignity in Animal Conservation 197

JOSEPH R. MENDELSON III

14 Seeing the Wild in Zoos by Seeing the Humans Too 213

AMANDA STRONZA

15 The Once and Future Rhino 229

MICHELLE NIJHUIS

Postscript: On Wildness and Responsibility 241

BEN A. MINTEER AND HARRY W. GREENE

Acknowledgments 247

List of Contributors 249

Index 255

A WILDER KINGDOM

1

ZOOS AND THE WILD: A RECONSIDERATION

BEN A. MINTEER AND HARRY W. GREENE

Zoos have always had, at best, a vexed relationship to what is considered the "real" wild. The celebrated writer and wilderness advocate Wallace Stegner, for example, saw them as representing the moral and cultural failure of nature conservation. "Something will have gone out of us as a people," he lamented in his influential "Wilderness Letter" of 1960, "if we ever let the remaining wilderness be destroyed; if we permit the last virgin forests to be turned into comic books and plastic cigarette cases; if we drive the few remaining members of the wild species into zoos or to extinction."[1] Stegner's ally, the prominent wilderness advocate David Brower, likewise remarked that zoos were little more than "prisons" for threatened wildlife and that a "death with dignity" for species teetering on the brink of extinction in the wild was preferable to putting them in zoos to try to recover them.[2] If Brower's philosophy had prevailed, we almost certainly would have lost his beloved California condor, an animal that was saved in no small part through the heroic efforts of zoos to breed and return the species to the southwestern skies.[3] Nevertheless, for ardent wilderness defenders like Stegner and Brower, the truth remained: the zoo was as far from "the wild" as you could get. In fact, it was where the wild went to die.

Although zoos have come a long way over the past few decades in terms of ecological aesthetics and design, this sense that zoos can only ever offer poor facsimiles of the wild has persisted. Even the largest, most immersive, and most naturalistic zoos, critics maintain, are unavoidably confined, contrived, and inauthentic environments in which any meaningful notion of the wild is simply unattainable.[4] Zoo animals, they point out, are under pervasive and intense human control: diet, veterinary care, reproduction, and so on are all carefully managed and choreographed, with natural phenomena like disease minimized and death kept mostly "backstage," hidden from public view. Also, despite their growing commitment to conservation and education, zoos are of course also entertainment facilities and so are incentivized to be responsive to visitors' aesthetic expectations and recreational preferences.

In his provocative book *A Different Nature*, David Hancocks, who led the redesign of the Woodland Park Zoo in the 1970s and later became the director of the Arizona-Sonora Desert Museum in Tucson (of that, more later), acknowledged a contradiction at the heart of the zoo idea. As he put it, "Zoos, from the most awful to the world's best, expose a perpetual dichotomy, which is the reverence that humans hold for Nature while simultaneously seeking to dominate it and smother its very wildness. . . . The wish to protect rare things is offset by a need for control."[5]

Yet could we envision a zoo where that wildness is not smothered but rather stoked? Or is the very notion of a "wilder zoo" an oxymoron? Maybe the more interesting and important question is whether zoos can meaningfully connect people to the wild *despite* their inevitable limitations. Regardless, what can we learn by putting our assumptions about the zoo and the wild in a more open dialogue, one where the goal is not to win an argument for or against zoos but to better understand, conserve, and hopefully coexist with other species, both in zoos and across the landscape?

WILD VISTAS

A Wilder Kingdom is our attempt to have this explicit conversation about the relationship between zoos and the wild and about the human place within them both. In the interest of full editorial disclosure, we should tell you that we've come to this discussion with a back story defined by a long-running debate between the two of us (but also, we're happy to add, by a mutual respect and friendship). Specifically, we've at times found ourselves split on how we understand the philosophical and moral constraints imposed by appeals to preserve some or another expression of the wild in conservation contexts.[6] One of us (Ben) has voiced concerns about the implications of aggressive interventions in nature for scientific and conservation ends, mostly out of a regard for maintaining a vision of the wild that, at least in some places and cases, isn't significantly under the human thumb.[7] The other (Harry) regards that last part as having always been unrealistic from ecological, evolutionary, and anthropological perspectives.

Despite our differences on these matters, we both agree that zoos present an especially fascinating case to consider within a broader inquiry into the wild, what it means, and what it might come to mean in the years ahead. They also provide an opportunity to reflect on how we should think of and relate to wild animals and places in a world that seems less wild each passing year. So we are united in this larger effort. But we also think it will be helpful in this section to disentangle our views on the wild and its connection to zoos and to discuss how our own interests and experiences have shaped our individual approaches. We'll partner up again in the final segment of this introduction with a brief outline of this book and a preview of the contributions of our distinguished collaborators.

Preservation and Pragmatism: Zoos and the Wild (Ben)

The relationship between zoos and the wild—and especially how we think about and value both—has been on my mind for some time. It figured into *The Ark and Beyond*, an earlier anthology I coedited on the past, present, and future of zoo conservation.[8] That book was a multidisciplinary volume exploring the history, philosophy, and practice of conservation in and by zoos as they move deeper into this part of their institutional mission. In my own contribution to the *Ark* collection, I grappled with the tension between zoos as entertainment centers presenting captive animals and our traditional ideas about and expectations for the wild and the wilderness as (very) lightly managed nature.

I've become absorbed by this zoo-wild relationship mostly because I find it stubborn to frame and difficult to think through. As an environmental ethicist with a deep interest in the American conservation tradition, it's proved challenging at times to square my sympathies for many of the values and goals of nature preservation (including, perhaps to Harry's chagrin, the Stegner-Brower variety) with some of the aesthetic and ethical trappings of the modern zoo. And yet my preservationist sympathies are also somewhat idiosyncratic.

The best way to put it is that while I strongly support the policy goals of nature preservation (e.g., stringent protection of endangered species, wilderness protection), I'm uncomfortable with some of the moral and political justifications marshaled to defend these ends. I'm especially wary of strident biocentric arguments that leave little room for the expression of other environmental values and visions beyond the embrace of nature's inviolability. And so for years I've defended an alternative outlook, a position I've dubbed "pragmatic preservation." It's admittedly an ungainly coinage, and so as far as I know it's remained solely in my possession. (In this I'm reminded of the remark by the philosopher Charles S. Peirce, who quipped in 1905 that his moniker for his ideas, "pragmaticism," was "ugly enough to be safe from kidnappers." He was right.) Pragmatic preservation as I see it simply means an approach to nature protection that recognizes the range of environmental values beyond intrinsic value (though it accepts a reasonable, i.e., nonideological, version of the latter), focuses on the practical consequences of ideas about nature rather than their purported theoretical "truth," and embraces an experimental and context-specific process of moral judgment about conservation efforts and policies rather than the prescription of fixed and universal moral principles.[9]

In any case, it became clear to me that there was an important discussion to be had about the possibilities and pitfalls facing zoos as many of them—though by no means all—seek to push deeper into an operational and educational philosophy that emphasizes a varying mix of naturalism, visitor immersion, and ecological design.[10] It was also obvious that this would be a conversation not just about how any "wilder zoo" might function in an animal-centered and ecological-evolutionary sense, e.g., by foregrounding a fuller complement of natural animal behaviors and relationships, from reproduction and competition to predation and even death. It would require reflection as well on how the public might be expected to perceive, value, and experience these markers of enhanced wildness in the zoo context. And all this would ultimately have a bearing on how we think about the meaning and value of nature preservation and the wild more generally. These questions and concerns quickly became my own personal inspiration for undertaking this journey into *A Wilder Kingdom*.

My perspective on this topic has been shaped by decades of thinking and writing in environmental ethics and conservation history, fields that regrettably have not done much to distinguish themselves in zoo studies. In these circles, considering zoos as emblems of and potential reservoirs for the wild and our experiences of it is a challenge, given the traditional stances most scholars in these fields have taken toward zoos and zoo animals over the years.

With few exceptions, environmental ethicists have paid scant attention to the normative questions raised by zoos, including their conservation activities and wider educational and public goals, especially when compared to the more traditional objects of nature protection like (in situ) endangered species protection and wilderness preservation.[11] Instead, most of the moral analysis of zoos has fallen outside environmentally focused ethics and has tended to dwell on the individualistic and often more salient animal welfare and rights issues relating to aspects of animal confinement, breeding, and display for human entertainment, as well as a general ethical scrutiny of the morality of animal captivity.[12]

Environmental philosophers who have considered the moral status of zoo animals vis-à-vis wild species and systems, such as Bryan Norton and Clare Palmer, have highlighted the neither-fish-nor-fowl nature of captive animals, arguing in different ways for adopting a context-specific understanding of our responsibilities to them in the zoo environment.[13] The mixed setting presented by zoos, where animals are not exactly wild yet also not tame, suggests a degree of moral responsibility for their care and well-being that goes beyond our obligations to their wild counterparts. Specifically, we have what moral philosophers would term a "positive" duty to ensure quality care of zoo animals, who are dependent on humans for their well-being. Our responsibility to wild animals, on the other hand, is often construed differently, i.e., as the "negative" duty to refrain from harming or interfering in the lives of individual animals. In short, we are obliged to let wild things be.

But it's not so simple. We may also have *positive* duties to animal populations in the wild given that human activities, from habitat destruction to climate change, have caused adverse changes in their ecological and evolutionary fortunes. Simply "letting wild things be" may therefore fail to uphold our obligations to them. In fact, significant interventions, such as augmenting animal populations through captive breeding or moving them to more suitable habitats to decrease their vulnerability to future climate change, may in fact be *required* as part of a responsible conservation ethic.[14] Similarly, we may have *negative* duties to zoo animals as well as positive. Refraining from activities such as intensive population-management protocols or animal shows may be required not only out of a respect for animal welfare and well-being but also as a broader responsibility to encourage their natural behaviors and relative wildness by reducing their manipulation, if not their ultimate dependence on human caretakers.

Zoos' mostly invisible status in environmental ethics parallels their relative neglect in the mainstream narratives of the wild in environmental history. You'll search in vain, for example, to find any significant reference to zoos in Roderick Nash's *Wilderness and the American Mind*, still the most influential and important account of human attitudes toward the wild ever written. Nash's book has certainly been a touchstone in my own thinking over the years. Interestingly, and also tellingly, Nash did reference zoos in voicing his concern that the administration of legally designated wilderness areas following the passage of the Wilderness Act in 1964 risked incarcerating the wild rather than letting it run free. "For all its benefits," he wrote, "the National Wilderness Preservation System might be regarded *as a kind of zoo for land*. Wilderness is exhibited in legislative cages, clearly mapped and neatly labeled. The unknown is known. Uncertainty decreases. So do risk and fear."[15] Once again, the implication is unmistakable: zoos are antithetical to the true wild.

Other than their role as foil for defenses of (what's said to be) the real wild, then, zoos rarely figure in the traditional histories of wildlife and wildland conservation. At most, they make cameo appearances in these accounts when they brush up against key episodes in the more familiar and canonical wildlife and wilderness story. So, for example, we may hear about the Bronx Zoo's William T. Hornaday and his pioneering efforts to captively breed and assist in the recovery of the American bison in the early twentieth century. Or zoos enter the picture as the places where some of the iconic species in the conservation narrative, from the passenger pigeon and the Carolina parakeet to the thylacine, made their last stands.[16]

On my reading, the contrast between zoos and what we typically think of as the "authentic" wild likely explains the indifference of most environmental ethicists and historians of wildlife and wilderness conservation. This attitude has no doubt been reinforced by older traditions in zoo design and exhibit craft that made the contrived nature of zoos blatantly obvious, from what the historian Simon Schama has described as the "cozy domesticity" of the London Zoo's early animal houses, modeled after an English village, to the austere bars and concrete of the "sanitary" design approach characterizing most American zoos from the 1950s well into the 1970s.[17] Far from a wild kingdom.

But it wasn't just this incongruence with ideals of the primitive wild that made the aesthetically discordant notes of the zoo so transparent. The zoo also clashed with the more manicured and managed version of nature captured in the early American (urban) parks movement. Frederick Law Olmsted, the celebrated landscape architect and planner who co-designed New York's Central Park, supposedly had little use for zoos, failing to include one in his original design for the park because he saw its artificiality as too dissonant with his vision of pastoral tranquility.[18] Olmsted's Central Park, engineered within an inch of its life in the middle of Manhattan, was far from a primeval wilderness, but a zoo was seen as beyond the pale, at least initially.[19]

A series of minirevolutions in zoo design beginning in the early twentieth century eventually ushered in more naturalistic design elements. Chief among them were the moated exhibits and landscape panoramas on display at the Tierpark zoo established by the German animal trader and early zoo impresario Carl Hagenbeck outside Hamburg in 1907.[20] In the 1920s and 1930s, zoos in Denver and St. Louis implemented the Hagenbeck style but tailored the exhibits' presentations of natural animal settings to local geology. This led to some unlikely animal-habitat pairings, such as the St. Louis Zoo's African Veldt exhibit, which mixed antelopes, giraffes, and zebras with a design featuring recreated boulders from Missouri rock formations.[21] And then there was the Bronx Zoo's immersive animal exhibits in the 1940s, which attempted to recreate exotic landscapes inspired by African habitats.[22] As the zoo historian Jeffrey Hyson notes, however, early zoo planners often made modifications to the new enclosures to ensure that the public could still have a direct view of the displayed animals, a reminder of the human interests at stake in even these "wilder" and seemingly more environmentally realistic displays.[23]

The redesign of the Woodland Park Zoo in Seattle in the late 1970s is commonly thought to have ushered in a more immersive and naturalistic design philosophy that would ripple throughout the zoo community in the ensuing decades (figures 1.1 and 1.2).

Although remnants of the bars-and-cage era remain in the American zoo today, they no longer dominate the zoo aesthetic and experience as they did a generation or two ago. And as I've written elsewhere, the emergence of attention-grabbing models outside the United States, such as Denmark's Zootopia, have pushed the commitment to immersion and (at least aesthetic) naturalism even further, proposing wall-less enclosures that creatively use water features and log piles to create barriers between animals and visitors. It's both a slick, twenty-first-century updating of

FIGURE 1.1. Animal cages at Woodland Park Zoo, 1916. Bear cage in foreground.

Source: Postcard collection (Record Series 9901-01), Seattle Municipal Archives/Wikimedia Commons.

FIGURE 1.2. A brown bear at Woodland Park Zoo, Seattle, WA, 2008.

Source: Photo credit: Benson Kua/flickr.

the older Hagenbeck design and a callback to his inveterate showmanship.[24]

These innovations and changes, however, haven't been able to outrun the deep-seated habit of viewing zoos and zoo animals as (at best) a disappointing approximation of the real, a "quasi wild." As the sociologist David Grazian puts it: "Nothing could be more self-evident than the staged or manufactured authenticity exhibited in zoo displays, no matter how naturalistic. No visitor approaching an Amur tiger's enclosure at the Philadelphia Zoo would somehow mistake its glass-walled habitat for the Siberian tundra. Yet as audiences we nevertheless expect zoo exhibits to prominently display landscapes that we at least associate with the natural environment, however imaginary and romanticized such renderings might be."[25]

Grazian is surely right about this. Yet our understanding of concepts like "natural" and "wild" as normative standards for judging zoo displays has become more complicated. For decades now, a swelling chorus of archaeologists, geographers, ecologists, environmental historians, and their allies has reminded us of the deep imprint of the cultural and the human in what we take to be natural and wild.[26] Add to that the recognition that most of the comparatively wild populations and places on Earth reflect some degree of human modification, manipulation, or management, and you come away thinking that the traditional notion of the wild as "pristine" and "untrammeled" no longer holds up to serious scrutiny, if it ever did. The emergence and impact of the idea of the Anthropocene, a construct acknowledging humans as a long-running environmental force on the planet, a species able to inscribe itself on the geological record, is just the latest and most expansive expression of this profound, perspective-altering insight.[27]

One conclusion, then, is that "the wild" can no longer be responsibly understood as a clear and simple antonym of "the cultivated" or "the managed."[28] Yet zoos still present a real challenge for thinking about the wildness—even in a qualified sense—of their animals and the zoo environment more generally. So maybe what we are left with, both in the zoo and in the ostensibly natural systems in which the wild counterparts of zoo animals roam, is a thousand points on a matrix encompassing the more- and less-intensively managed, rather than the binary of wild *or* domesticated. In this understanding, which accepts a more graded, patchy view of the wild and its forms across the landscape, from rugged and far-flung wildlands where people are scarce to the heart of the urban core, zoos and the animals within them can be more or less wild depending on the biological, behavioral, and custodial circumstances in which they dwell and act, not to mention their cultural context.

Thinking more pluralistically about the wild and its variable forms and intensities could also help us avoid getting further mired in endless philosophical and political debates over the value and place of "the wild" in the conservation agenda, debates where the wild is often construed (by both defenders and critics) as some sort of Platonic ideal. As Curt Meine and others have reminded us, there is no pure or absolute wild; depending on how we stipulate the meaning of "wild," there are only *more-* and *less-*wild places, behaviors, and conditions.[29] In short, the meaning and value of the wild is inexorably relative and always culturally and context bound. Yet saying as much doesn't mean that we should weaken our commitment to saving wild species and wildlands from destruction or that we should tear up the 1964 Wilderness Act and purge all mention of the wild and wilderness from our conservation goals and politics.[30] It just means we need to recognize that these values and agendas exist within a wider spectrum of environmental thought and action.

The same goes for our ways of moralizing about "the zoo." Whenever I get into a conversation with someone about zoos and mention that I'm an environmental ethicist who has lately thought and written a lot about them, they almost always ask me the same question: "So do you think zoos are good or bad?" I answer that I don't believe zoos are either inherently *only* "good" or inherently *only* "bad." Instead, I say that there are better and worse zoos when it comes to animal care, exhibit design, and the quality of their education, research, and field conservation programs. To some, this is an acceptable response. To others, it's a pitiful dodge.

But to me, the "it depends" view is a more difficult and demanding position to hold than simply saying zoos "yea" or "nay." It makes us think more locally, concretely—again, contextually—about the specific operating conditions, priorities, and conservation efforts of particular zoological facilities rather than passing blanket judgment on a generic abstraction. Obviously, this response won't satisfy those who view the very *idea* of a zoo as anathema to any responsible animal ethic. Even the largest, most naturalistic, and most professionally run zoos are still seen as violating the freedom and dignity of animals by holding them in captivity and putting them on public display.[31] To these critics, my mealy-mouthed qualifications about better and worse zoos are equivocations. Or even more objectionable, a feckless apologia for what they see as a thoroughly irredeemable institution.

I understand their position. But it's also why we need to tell a more nuanced story about the wild "out there" and to think more pragmatically about representing wildness and our experience of it in highly managed settings like zoos. Many of the doctrinaire anti-zoo arguments (e.g., artificial prisons for wildlife, etc.) simply land differently when the familiar contrastive ideal, i.e., the wild, is viewed less dogmatically and, especially, less romantically.

Even David Hancocks, who has persuasively documented the zoo's shortcomings, sees their potential to reorient the way we value nature and the wild. "With a few changes," he suggests, "we can design zoos that convey the richness of the natural world and that carry vital messages about our need to love, care for, and protect its diversity. . . . Zoos have the capacity to help us refocus our views of wild animals and wild places. They can encourage a new understanding of Nature."[32]

Hancocks's words serve as an apt point of embarkation for *A Wilder Kingdom.* I'm fortunate to be joined in this voyage by my coeditor, Harry Greene, who has challenged and inspired my own thinking (often at the same time) about the richness of the natural world and our responsibility to understand, care about, and protect it. A key feature of Harry's ethos is the respectful but active participation in wildlife populations and wildlands. It is a vision that has intriguing implications for understanding zoos as spaces where people can meaningfully connect to wild animals and places, even if in only representational ways.

More Real Than Imagined? (Harry)

I grew up a naturalist in the preservationist tradition, only to have it challenged by decades of field teaching and research on four continents, certain Indigenous perspectives, and becoming a late-onset deer hunter. A place is wilder, I now believe, for housing locale-appropriate ecological and evolutionary diversity—from microbes to large herbivores and predators—instead of the absence of human impact; likewise, organisms are wilder when engaged in ecological and evolutionary processes, rather than for not interacting with people. My piece on rewilding captivity in an earlier book on zoos that Ben coedited focused on carcass feeding, culling, and other maintenance practices, whereas here I'll argue that what zoos might and should be hangs in part on the very meaning of *wildness.* My personal account of grappling with that word seeks to clarify what it might signify for zoos as well as our relationships with other species more generally.[33]

In the early 1990s, while sifting natural history literature for a book on snakes,[34] I realized that lemurs, monkeys, and apes kill and are killed by my favorite reptiles. Serpents dining on mammals is evolutionarily older than primates, which implies tens of millions of years of bad blood between our lineages—yet I'd built a career out of studying and defending limbless scaly creatures, preaching that even venomous kinds are easily avoided. Now what was I to make of this ancient baggage, in terms of encouraging humans to appreciate snakes?

Fast forward to 2008, when I cotaught behavioral ecology to Cornell undergraduates at a Kenya site inhabited by big cats, hippos, buffalo, and elephants. While we professors supervised projects on ants, geckos, and trees, our field-station staff associate attended to wind direction, scrutinized thickets for natural-born killers, and kept lookout for anyone who might gore or squash us. Another eye-opener, thanks to the antelope, zebras, and other herbivores, was the volume and variety of dung. Everything from dainty pellets to fibrous cannonballs littered the ground, even around water holes, and we saw how amid a flourishing megafauna, despite the usual first-world aversions to excrement, poop, too, can be pristine.

Beyond danger and droppings, however, my most nagging revelation in Kenya was that idealizing wildness and wilderness as a state wherein people do not affect other species conflicts with the facts of natural history. On foot among all those large mammals, guided by the critter-savvy descendant of local pastoralists, I first grasped that assigning our ancestral continent no human influence would require going back two million years. A strict preservationist, I now saw, could only diagnose Africa as *wild* before we evolved from the fossil Lucy and her kin. A cornerstone tenet for many environmentalists thus is irrelevant on the landmass of our origin.

This discontent worsened after returning to New York. Soon I surmised that even in the hemisphere of my birth, the benchmark for no human impact must be before ancestral Native Americans colonized from Asia, perhaps more than twenty thousand years ago. Inspired by teaching among big mammals in Kenya and thereafter by hunting deer for food in Texas, I questioned *wilderness* defined as a place where Indigenous people are dismissed as ecologically and evolutionarily irrelevant. Might such thinking, at odds with reality as I newly perceived it, impede conservation? Could such romantic, arguably racist attitudes be problematic for saving the dangerous and otherwise unpopular animals that have captivated me since childhood?

I'd made it through midlife comfortable with beliefs that now seemed misguided, yet as a newly born-again predator I felt more engaged with nature than ever before. In the few minutes it took to field-dress that first slain doe—slicing her open, wet viscera steaming on frozen ground, bloody hands trembling and heart pounding—I'd gone from spectator to uneasy participant. Moreover, this was in the Texas Hill Country, a place with which I've been familiar since childhood. How could I reconcile these tensions within an acutely personal worldview? Can anything novel be said about hunting, wildness, and wilderness values, given that volumes already have addressed those topics? And why should connecting them anew matter to nature lovers or anyone else?

Back to serpents, we have killed and been killed by them since the Plio-Pleistocene and much earlier. Constrictors, key players in early snake evolution, likely have influenced our genes and culture since the very dawn of team primates, roughly ten times longer ago than our seven-million-year-old divergence from bonobos and chimps. Fast forward from that split to Stone Age hominins who butchered African pythons and likely were preyed upon by them, thence two million more to the mid-twentieth-century Agta of the Philippines, who ate pythons, were eaten by the giant reptiles, and hunted macaques, deer, and pigs. For about 75 million years, then, primates and snakes have influenced each other as predators, prey, and competitors. The Agta and surely much earlier hominins were aware of all three relationships. Likewise, the origin of snake venoms and fangs more than 35 million years ago perhaps favored the invention of weapons by capuchins, chimps, and our own lineage, whereas upright hominins wielding clubs and projectiles might have led to evolution of long-distance "spitting" by African and Asian cobras.[35]

With serpents, then, it's been an arms race spun out over geological eras, pitting their increasingly specialized teeth, toxins, and defensive behavior against our ever more sophisticated vision, manual dexterity, cognitive abilities, and culture. Moreover, snakes are just one among many organisms with whom we primates long have had complex, often mortal interactions. Two million years ago, giant raptor talons pierced the eye sockets of a hominin who became the first described fossil of Lucy's *Australopithecus* lineage; today, eagles prey upon monkeys, even rarely attack human children. Turning the tables, we have hunted birds and mammals with falcons, hawks, and eagles for at least ten thousand years. Our cooperation with dogs is older still and paralleled in simple form by grass-eating geladas, Ethiopian baboon relatives that allow wolves to forage among them and consume rodents with whom they compete for food.[36]

Lurking among hunting and other ecological legacies are implications for modern dilemmas, like eating meat and coexisting with dangerous animals, all part and parcel of rethinking wildness and wilderness values. Mortality, the other side of life's coin, recurs as a motif therein, filtered in my case through childhood stints on grandpa's Texas dirt farm, then firsthand experience with violence and death as a medic and soldier, a scientist focused on predators, and a meat hunter. And beyond our individual and cultural contexts, meanings can be warped rhetorically—by folks espousing our paleo past to justify slaughtering feedlot cattle, for instance, or by the anthropologist Matt Cartmill, in *A View to a Death in the Morning*, to push an antihunting agenda.[37]

Although no one stumbles over calling snakes or eagles hunters, word choices are especially important in the context of our complex biocultural history. That said, killing characterizes hunting overall, not specific tactics, implements, goals, attitudes, or outcomes. Moreover, our oldest, most widespread motivation for this violent act has been to eat meat, but we likely also have trapped carnivores for fur about twenty-five thousand years and sewn animal hides into clothing for most of our 300,000-year-long history as a species. Hunting and trapping tie into wildness because benefiting from the bodies of other organisms always has been among the most common ways that humans engage in natural processes (others include defecating and dying). I believe, however, that participation more generally—not specifically killing—is a key topic for understanding wildness.

Central to discussions of wildness and wilderness values are two mindsets, contrasted in the early twentieth century by what are now termed *instrumental* (utility-based) versus *intrinsic* (self-worthy) values. Instrumentalist *conservationism* was about "wise use" of nature for the good of people and proved too one-sidedly selfish for me and many other outdoor lovers. *Preservationism* instead celebrated aesthetic and spiritual rewards and was embraced mainly by North Americans of European descent, usually men, at least historically. Preservationists write as if humans are not part of nature, although they also complain about our being "apart." Wild creatures are self-willed rather than influenced by us, as if those were exclusive options. Most crucially, wild places, especially those labeled wilderness, are imagined as untrammeled, wherein we take only photos and leave only footprints. Wilderness is an intrinsically valuable landscape, a cathedral in which those with means and access can worship. Wilderness is not a place wherein humans obtain food or otherwise participate in biologically meaningful ways.

Today, *self-willed* and *untrammeled* as diagnostics are still defended with religious-like fervor in some circles. Kenneth Brower, in an essay entitled "Leave Wilderness Alone," stereotyped challengers as "an Ivy Rebellion of pale men in tweed." He recalled "following my father [wilderness advocate David Brower, mentioned earlier] up high passes over which the Paiutes traded . . . sites where they paused to pressure-flake obsidian arrowheads . . . [and even as a] small boy . . . [I] distinguish[ed] half-buried artifacts from, say, a Sierra ski resort with its lifts rumbling or a six-lane highway with exits to malls." Then, as if realizing just *one* aspect of material culture—isolated among childhood keepsakes from any ecological or cultural context—might not *quite* do justice to Indigenous lives and environmental relationships, Brower adds, "Yes, *some* tribes set fires to keep *meadows* open . . . the earliest Amerindians *may have helped ease* the Pleistocene megafauna to extinction. But how can these aboriginal impacts be compared, *as deconstructionists routinely do*, to transformation of the planetary surface and atmosphere wrought by technological civilization?"[38]

Minimization of earlier human involvement with nature is extensively contradicted by history. For centuries on the southern Great Plains, Comanche and Kiowa ruled over tens if not hundreds of thousands of horses; their war shields portrayed longhorns as emblematic of power, equivalents of bison in that regard, not as European livestock. California's Channel Island foxes diverged from a mainland species roughly seven thousand years ago, apparently having been brought over in boats by the Chumash. And reaching farther back in time, palms are prominent today in "pristine" Amazonian rainforests because people used them for food and shelter more than ten thousand years ago; Andean squash and manioc were brought there about the same time, also with landscape-scale ecological effects. Anthropogenic fires have shaped forests and grasslands for millennia, and we likely have long affected other species—almost two million years before the famously controversial late-Pleistocene megafaunal extinctions, a guild of middle-sized African carnivores disappeared right about when and where hominin brains became rapidly enlarged.

Beyond debates about whether we are part of nature, the philosophical underpinnings of wildness have been muddied by framing values as instrumental *or* intrinsic. Consider that the conservation icon Michael Soulé derived his ethics from Buddhism, but for those of us who aren't religious yet adore sun-splashed canyons and vermillion flycatchers, whence could come noninstrumental values other than as norms within what philosopher Joshua Greene calls our "moral tribes"? Ben may part with me here, but are not intrinsic values *always* faith based, and, if so, which faith offers the *only* correct worldview? To use a familiar case in point, some cultures vilify serpents, others deify them, and who's to say whose beliefs must hold for everybody? The dichotomy further suffers from ignoring geography, religion, economic status, and other factors that shape values; compassion and empathy affect moral choices too, yet they can be biased by familiarity, size, and danger. Withal, these complexities underscore roles for candor, clarity, and humility as we navigate ethical dilemmas—just the blend of pluralism and pragmatism promoted in Ben's writings. After all, Albert Schweitzer, widely celebrated for "revering all life" during his medical missionary work in early twentieth-century Africa, kept a rifle at hand for killing snakes. And Michael Soulé was an elk hunter.[39]

A key point here is that "wild" isn't synonymous with "peaceful," no matter the odd claim of John Muir, founder of the Sierra Club, never to have seen a drop of blood in all his wanderings. Yes, wild places are enlivened by melodious insects, frogs, and birds, but they also house the bellowing rivalries and thundering hooves of giant grazers and browsers, the bone-crunching and snarling of predators, the moaning of prey. Out on the Serengeti, lion cubs succumb to infanticide by males newly joining a pride; impala lambs are killed clumsily by young cheetahs perfecting skills required later in life. And Albert Schweitzer persecuted my beloved snakes because back then, where he lived, they were a horrific threat to life and limb. Meanwhile, in Muir's California, snow-capped peaks, crystalline streams, and cerulean skies, celebrated for tranquility, also frame the putrid-sweet stench and buzzing flies around a mule deer carcass—reality checks on alpine flowers whose idyllic colors and odors facilitate pollinator-assisted sex, on fruits whose zesty hues and flavors have been selected for seed dispersal rather than enhancing our breakfasts.

These and countless other natural history details confirm that wildness, even more real than imagined, isn't tidy. Can wilderness values be otherwise, then, unless premised on flawed distinctions between instrumental and intrinsic values, domesticated and wild species, or on notions of humans as separate from nature? As backdrop for a realistic alternative vision, paraphrasing Evelyn Hutchinson, cycles of life and death amount to evolutionary plays in ecological theaters. Their plotlines unfold among organisms aboard an ever-changing Earth, so viewing us as actors rather than audience can inspire empathy for entire ecosystems and life's four-billion-year history.[40]

As with microbes, plants, fungi, and other animals, we receive from, provide for, and cooperate with other species; like them we partake in energy flow, cycle water and nutrients. Other organisms influence our fates and evolution, just as we affect theirs. And as humans we view it all through biocultural lenses and individual experiences. For this grandson of a subsistence farmer-hunter, these truths have felt most poignant when stakes were on the higher side—within killing distance of a deer or boar whose flesh would sustain me, while being shadowed by a puma or a jaguar who perhaps wanted to eat me, or when hiking among buffalo and elephants who in self-defense might gore or trample me—or even more existentially, while contemplating cycles of decomposition and renewal, like when insects transform my own droppings into soil and vultures scavenge leftovers from game I kill for food. Those intimate relationships epitomize what the Rarámuri scholar Enrique Salmón calls our *reciprocal ecological kinships* or *eco-kin-centricity*, and they can inspire both scientific and Indigenous perspectives on nature.[41]

Habitats with dung beetles are wilder for hosting those little waste management engineers, but more's at stake than ecosystem empathy and evolutionary heritage. Henry David Thoreau famously wrote that "in wildness is the preservation of the world." Like him and countless others, out in nature I experience a nuanced, sublime reverence; embedded in wildness, I, too, achieve a special humility, both bracing and calming, one that encompasses dependence and reciprocity within schemes far grander than individual lives. So yes, wildness inspires emotional resonance, expressions of which by Muir, Thoreau, and other preservationists are admirable. Natural history teaches us, however, that wildness comes in diverse flavors, including places wherein humans are participants rather than spectators. Wildness, as always has been true in our African homeland, is thus realistically defined by human coexistence with the likes of lions and elephants, not by our absence for its own sake. In my reformist view, then, full-tilt wildness entails humans *mindfully engaging* with soils, plants, vegetation, the largest locale-appropriate herbivores and predators, and those innumerable smaller organisms who tie us all together. A corollary is that newcomers to a place likely have much to learn from those, Indigenous and otherwise, who have much older relationships with local landscapes.

Of course, wildness and wilderness values *are* threatened in a world overstuffed with people, and as we alter habitats, large and otherwise inconvenient species are the first to disappear. Worse yet, devils prowl the details. Ever more of us demand minimally risky lives, yet this becomes problematic when activists impose privileged values on people who live with dangerous species and other threats to their well-being. Tolerance for serpents might more likely come, for example, with eliminating many of the 100,000 global annual snakebite deaths, most in less-developed countries. Likewise, most suburbanites no longer hunt for food and clothing nor grow and gather vegetables or have decomposers directly transform their wastes—but might we look to those who do for inspiration and collaborations, rather than patronize them or worse? Maximal wildness requires low human population densities, and thus many of our lives score moderately to very low across a span of possibilities. My bet, nonetheless, is that we could do better, live more wildly even within urban centers, and there, too, gods lurk among the details. Does anyone deny cities with stooping peregrines and howling coyotes are wilder than those without them? And what about wilderness values within a scientifically informed, eco-kin-centric worldview?

A few years after teaching in Kenya and becoming a hunter, I hiked off-road for several days in South Africa's Kruger National Park. As our little group stepped out from some trees, having just admired a lion's tracks, a rhinoceros was a dozen or so yards ahead, most of his three-ton bulk submerged in a wallow. His ears were erect and twitching, his sense of smell surely on high alert. Moments later, this ponderous creature lurched up out of the mud with breathtaking agility, faced us for a few long seconds, and sauntered off into nearby grasslands. Time never truly stands still, but his colossal, vibrant, inscrutable *otherness* was stunning. If birds sang, I didn't hear them. I could scarcely think, let alone speak. The rhino's black-dot eyes were riveting, and beyond the surpassing puzzle of what was going on in that big blocky noggin, three things haunt me from our time together: that he might have defensively charged had not a preferred escape route remained open, thanks to the wisdom of our park ranger guides; that my life has been enriched by him and his colossal kin; and that humanity will be impoverished if they disappear.

As if to emphasize grim prospects for the remaining megafauna and maximum wildness, on that same afternoon we walked up on the moldering carcass of another rhino, slaughtered by poachers for her horn. Then years later, viewing black contour drawings of woolly rhinoceroses in a French cavern and recalling the Kruger animals, I was astonished to realize that Pleistocene artists with torches had crouch-crawled a mile underground to portray the great beasts. And as I reflected upon those ancient people's animistic worldviews, so manifestly shaped by plant-eating behemoths and the occasional cave bear or lion, I wondered if photos of rhinos might soon provide eulogies for the last five species of that forty-million-year-old "nose-horned" lineage? How utterly heartbreaking if they blink out thanks to superabundant, greedy, and cruel humans—an entire guild of eco-kinfolk, present throughout our own piddling two-million-year history, vanquished from Earth's stage. How emblematic of the challenges faced today by wild places, wild animals, and those of us who love them.

It happens that my journey has converged on Ben's with respect to some open-ended questions for this volume. Could zoos be wilder if we at least simulate participation in ecological and evolutionary processes? Can explicating those wilder relationships for zoo visitors—or better yet, *drawing them in*—inspire conservation of wilder places? And perhaps most fundamental, how might biological, anthropological, and eco-kin-centric considerations of wildness influence moral judgments about nonhuman animals, captive and otherwise?

PROMPTS, PROVOCATIONS, AND PLACE

In this book we've hand-picked a distinguished and diverse roster of thinkers to explore these and related ideas and issues in greater depth and dig more deeply into this complex intersection of zoos and the wild. The contributors include writers with special expertise in zoos and human-wildlife relationships but also those who have made their mark in other fields and genres (e.g., evolutionary ecology, ethology, anthropology, environmental history, science journalism, nature writing) and who approach the question of the wild, the zoo, and their intersection from new and different angles. We prepared a set of questions, motivated by many of our reflections in this chapter, to get everyone thinking.

Specifically, we invited them to explore the meaning of "wild" and "wildness" across the landscape and to consider how different perspectives—scientific, psychological, aesthetic, cultural, and so on—shape our understanding and experience of zoos and the wild populations and places beyond them. Our intent was to put down a series of preliminary provocations and ask each author to pick up the ones that spoke to them. As you'll see, it's an approach that has resulted in a lively and engaging series of reflections. The narrative begins with a series of broadly thematic essays exploring our shifting ideas about the wild and its expression in zoos before moving on to more specific discussions of some of the many scientific, design, and operational implications of a wilder zoo. The closing essays then look outward, exploring the connections between zoos, the wild, and their surrounding human and ecological communities.

A Wilder Kingdom is grounded in a particular and very special place: the Arizona-Sonora Desert Museum (ASDM) located on the outskirts of Tucson, Arizona. Established in 1952, ASDM is an innovative and unusual hybrid of zoo, aquarium, botanical garden, natural and cultural history museum, art gallery, and nature center.

Adopting a regional, place-based approach to its exhibits, the Desert Museum displays only species native to the Sonoran Desert while carrying a deep commitment to conservation, education, and community engagement. It's also one of the most naturalistic and immersive zoos in the country, a status facilitated by its being surrounded by miles of protected desert lands, a geographical context that works to blur the boundary between the zoo and the landscape beyond the perimeter. As we see it, the ASDM suggests one possible future in which wildness and naturalism are core features of zoo design and visitor experience, especially with respect to conservation education. We were delighted, therefore, when the ASDM agreed to host a gathering of our authors in the fall of 2021, where we all shared earlier versions of the ideas and themes corralled in this book. The ASDM environs and vision have provided our project with a vital anchoring and tangible reminder of the real-world implications of these ruminations on zoos and the wild.

Taken as a whole, we think the chapters that follow underscore how our ways of enclosing and displaying nature in zoos are neither as fixed nor as inevitable as is often assumed. Just as the older bars and cages eventually gave way to a more naturalistic zoo tableau, our traditional view of the wild as the antithesis of the zoo seems to be evolving into a more nuanced and diverse set of understandings and images. We hope you'll find the discussions in the pages that follow as informative, challenging, and thought-provoking as we have.

NOTES

1. Wallace Stegner, "Wilderness Letter," (1960), in *The Sound of Mountain Water* (New York: Vintage/Penguin, 2017), 140.
2. David Brower, quoted in Noel F. R. Snyder and Helen Snyder, *The California Condor: A Saga of Natural History and Conservation* (San Diego, CA: Academic Press, 2000), 130.

3. Peter S. Alagona, *After the Grizzly: Endangered Species and the Politics of Place in California* (Berkeley: University of California Press, 2013); see also John Farnsworth, "The Condor Question Revisited," *Minding Nature* 8 (2015): 31–36, https://www.humansandnature.org/filebin/pdf/minding_nature/may_2015/TheCondorQuestion.pdf.
4. See, for example, Marc Bekoff and Jessica Pierce, *The Animal's Agenda: Freedom, Compassion, and Coexistence in the Modern Age* (Boston: Beacon); and Emma Marris, *Wild Souls, Freedom, and Flourishing* (New York: Bloomsbury, 2021).
5. David Hancocks, *A Different Nature: The Paradoxical World of Zoos and Their Uncertain Future* (Berkeley: University of California Press, 2001), 6.
6. An illustration of our differences on these questions can be found in our articles for the Center for Humans and Nature series How Far Should We Go to Bring Back Lost Species? (https://www.humansandnature.org/how-far-should-we-go-to-bring-back-lost-species): Ben Minteer, "Lost Species Should Stay Extinct," Center for Humans and Nature, December 1, 2014, https://humansandnature.org/extinct-species-should-stay-extinct/; Harry W. Greene, "As Far as We Can Go, as Far as We Want to Go . . .," Center for Humans and Nature, December 1, 2014, https://humansandnature.org/as-far-as-we-can-go-as-far-as-we-want-to-go/.
7. Ben A. Minteer, *The Fall of the Wild: Extinction: De-Extinction, and the Ethics of Conservation* (New York: Columbia University Press, 2018).
8. Ben A. Minteer, Jane Maienschein, and James P. Collins, eds., *The Ark and Beyond: The Evolution of Zoo and Aquarium Conservation* (Chicago: University of Chicago Press, 2018).
9. Minteer, *Fall of the Wild*. See also Ben A. Minteer, *Refounding Environmental Ethics: Pragmatism, Principle, and Policy* (Philadelphia: Temple University Press, 2011).
10. This trend is described in Hancocks, *A Different Nature*; Elizabeth Hanson, *Animal Attractions: Nature on Display in American Zoos* (Princeton, NJ: Princeton University Press, 2002); and David Grazian, *American Zoo: A Sociological Safari* (Princeton, NJ: Princeton University Press, 2015).
11. Exceptions are discussions in Bryan G. Norton, Michael Hutchins, Elizabeth F. Stevens, and Terry L. Maple, eds., *Ethics on the Ark: Zoos, Animal Welfare, and Wildlife Conservation* (Washington, DC: Smithsonian Institution Press, 1995); Jozef Keulartz, "Captivity for Conservation? Zoos at

a Crossroads," *Journal of Agricultural and Environmental Ethics* 28 (2015): 335–51; and the relevant essays in Minteer et al., *The Ark and Beyond.*

12. E.g., Dale Jamieson, "Against Zoos," in *In Defense of Animals*, ed. Peter Singer (Oxford: Blackwell, 1985), 108–17; and Dale Jamieson, "Zoos Revisited," in *Ethics on the Ark*, ed. Norton et al., 52–66. See also Lori Gruen, ed. *The Ethics of Captivity* (Oxford: Oxford University Press, 2014).
13. Bryan G. Norton, "Caring for Nature: A Broader Look at Animal Stewardship," *Ethics on the Ark*, ed. in Norton et al., 102–21; and Clare Palmer, *Animal Ethics in Context* (New York: Columbia University Press, 2010).
14. Ben A. Minteer and James P. Collins, "Species Conservation, Rapid Environmental Change, and Ecological Ethics," *Nature Education Knowledge* 3 (2012): 14, https://www.nature.com/scitable/knowledge/library/species-conservation-rapid-environmental-change-and-ecological-67648942/.
15. Roderick Frazier Nash, *Wilderness and the American Mind*, 5th ed. (New Haven, CT: Yale University Press, 2014), 339; emphasis added.
16. Although wilderness and conservation historians have paid little attention to zoos, historians of animals and the life sciences have done much better on this score. See especially Nigel Rothfels, *Savages and Beasts: The Birth of the Modern Zoo* (Baltimore, MD: Johns Hopkins University Press, 2002); Mark V. Barrow Jr., *Nature's Ghosts: Confronting Extinction from the Age of Jefferson to the Age of Ecology* (Princeton, NJ: Princeton University Press, 2009); and Daniel E. Bender, *The Animal Game: Searching for Wildness at the American Zoo* (Cambridge, MA: Harvard University Press, 2016).
17. Simon Schama, *Landscape and Memory* (New York: Vintage, 1995), 563. The evolution of zoo design is discussed in Jeffrey Hyson, "Jungles of Eden: The Design of American Zoos," in *Environmentalism in Landscape Architecture*, ed. Michel Conan (Washington, DC: Dumbarton Oaks Research Library Collection, 2000); and Hancocks, *A Different Nature.*
18. Hyson, "Jungles of Eden."
19. A zoo was eventually included in the park design at the behest of park commissioners. NYC Parks, "History of Central Park Zoos," https://www.nycgovparks.org/about/history/zoos/central-park-zoo.
20. The best discussion of Hagenbeck's zoo philosophy and vision remains Rothfels, *Savages and Beasts.*

21. See Hanson, *Animal Attractions*, 150.
22. Gregg Mitman, "When Nature Is the Zoo: Vision and Power in the Art and Science of Natural History," *Osiris* 11 (1996): 117–43; see also Bender, *The Animal Game.*
23. Hyson, "Jungles of Eden."
24. Minteer, *Fall of the Wild*, chap. 3.
25. Grazian, *American Zoo*, 18.
26. E.g., William M. Denevan, "The Pristine Myth: The Landscape of the Americas in 1492," *Annals of the Association of American Geographers* 82 (1992): 369–85; William Cronon, "The Trouble with Wilderness, or, Getting Back to the Wrong Nature," *Uncommon Ground: Toward Reinventing Nature*, ed. in William Cronon (New York: Norton, 1995), 69–90; Mark David Spence, *Dispossessing the Wilderness: Indian Removal and the Making of the National Parks* (New York: Oxford University Press, 1999); and Erle C. Ellis, "Anthropogenic Transformation of the Terrestrial Biosphere," *Philosophical Transactions of the Royal Society A* 369 (2010): 1010–35.
27. See, e.g., Ben A. Minteer and Stephen J. Pyne, eds., *After Preservation: Saving American Nature in the Age of Humans* (Chicago: University of Chicago Press, 2015); and Simon L. Lewis and Mark A. Maslin, *The Human Planet: How We Created the Anthropocene* (New Haven, CT: Yale University Press, 2018).
28. Curt Meine, "A Letter to the Editors: In Defense of the Relative Wild," in *After Preservation*, ed. Minteer and Pyne, 84–95.
29. Meine, "A Letter to the Editors." See also the fine collection assembled in Gavin Van Horn and John Hausdoerffer, eds., *Wildness: Relations of People and Place* (Chicago: University of Chicago Press, 2017).
30. See, for example, the claims advanced by Peter Kareiva, Michelle Marvier, and Robert Lalasz, "Conservation in the Anthropocene. Beyond Solitude and Fragility," *Breakthrough*, Winter 2012, https://thebreakthrough.org/index.php/journal/past-issues/issue-2/conservation-in-the-anthropocene.
31. Jamieson, "Zoos Revisited"; and Marris, *Wild Souls.*
32. Hancocks, *A Different Nature*, xv.
33. Harry W. Greene, "Re-Wilding the Lifeboats," in *The Ark and Beyond: The Evolution of Zoo and Aquarium Conservation*, ed. Ben A. Minteer, Jane

Maeienschein, and James P. Collins (Chicago: University of Chicago Press, 2018), 360–69. Also, small portions of my introduction here are revised from earlier explorations of these topics, in which I supplied more extensive references: Harry W. Greene, *Tracks and Shadows: Field Biology as Art* (Berkeley: University of California Press, 2013); Harry W. Greene, "Rewilding Our Lives," *Mind and Nature* 8 (2015): 18–24, https://humansandnature.org/rewilding-our-lives/; and Harry W. Greene, "Pomegranates, Peccaries, and Love," *Ecopsychology* 12 (2020): 166–72.

34. Harry W. Greene, *Snakes: The Evolution of Mystery in Nature* (Berkeley: University of California Press, 1997).
35. Thomas N. Headland and H. W. Greene, "Hunter-Gatherers and Other Primates as Prey, Predators, and Competitors of Snakes," *PNAS* 108 (2011): 20865–66, E1470–74; Taline Kazandjian et al., "Convergent Evolution of Pain-Inducing Defensive Venom Components in Spitting Cobras," *Science* 371 (2021): 386–90.
36. Lee R. Berger, "Predatory Bird Damage to the Taung Type-Skull of *Australopithecus africanus* Dart 1925," *American Journal of Physical Anthropology* 131 (2006): 166–68; Vivek Venkataraman et al., "Solitary Ethiopian Wolves Increase Predation Success on Rodents When Among Grazing Gelada Monkey Herds," *Journal of Mammalogy* 96 (2015): 129–37.
37. Matt Cartmill, *A View to a Death in the Morning: Hunting and Nature Through History*, (Cambridge, MA: Harvard University Press, 1993).
38. Kenneth Brower, "Leave Wilderness Alone," *Outside Magazine*, October 13, 2014, https://www.outsideonline.com/outdoor-adventure/environment/leave-wilderness-alone/; my emphasis.
39. Joshua Greene, *Moral Tribes: Emotion, Reason, and the Gap Between Us and Them* (New York: Atlantic, 2014).
40. G. Evelyn Hutchinson, *The Ecological Theater and the Evolutionary Play* (New Haven, CT: Yale University Press, 1965).
41. Enrique Salmón, "Kincentric Ecology: Indigenous Perceptions of the Human-Nature Relationship," *Ecological Applications* 10 (2000): 1327–32; Enrique Salmón, *Eating the Landscape: American Indian Stories of Food, Identity, and Resilience* (Tucson: University of Arizona Press, 2012).

2

BETWEEN WORLDS

A Conversation Among the Cranes

CURT MEINE

Every spring and fall a family conversation echoes through the hills, wetlands, and farm fields of south-central Wisconsin. Opening remarks come from above. Migrating sandhill cranes call from the skies over the headquarters of the nonprofit International Crane Foundation (ICF), just north of the city of Baraboo. The calls prompt other cranes, earthbound, to raise their voices. As the sandhills bend their path and continue the discussion, excitement spreads like a rumor from crane pen to crane pen. The heads of a hundred and twenty cranes lift, long necks draw back, and the bugles cut loose. The clamor rattles the fences and calls human neighbors to attention. Even the resident crane biologists, caretakers, and educators, who know these voices well, pause to take note.

This conversation would not occur under normal circumstances. And it can take place only here. ICF is the only point on the planet where all fifteen members of the Gruidae family are present. In its capacity as the global center for crane conservation, ICF houses representatives of all the world's crane species. From their five home continents—only Antarctica and South America lack cranes—these birds have been assembled on this unassuming patch of the American Midwest to answer a special calling. Some of the cranes inhabit ICF's public facilities, where the foundation's 25,000 annual visitors encounter them, hear their stories, and learn about their biology, ecology, and conservation needs. Others reside away from the public in "Crane City," ICF's captive breeding facility, where they may be propagated for reintroduction, research, and education programs. (Over the years, active breeding efforts have come to focus mainly on the two most critically endangered species, the whooping crane and Siberian crane.)

The discerning human ear can make out the family's varied voices amid the chorus of cranes. The gray and black crowned cranes utter short, guttural honks. Those with more elaborate, elongated tracheas—the whooping crane, red-crowned crane, and sarus crane—announce themselves in bright, stentorian tones. Siberian and wattled cranes occupy the higher register. And above the din the wild sandhill cranes issue their coarse cries.

The human eavesdropper on all the crane-talk may also pick up undertones of irony. The cranes, having spent eons diversifying from their shared ancestor and following their own adventurous evolutionary paths, have been reunited by their increasing vulnerability. Cranes are among the most endangered families of birds in the world. Ten of the fifteen species are included on the IUCN Red List of Threatened Species. Another is listed as near threatened. Subspecies and populations of even the more common species are at risk. One taxon, the Philippine subspecies of Sarus crane, *Grus antigone luzonica*, has not been recorded since the late 1960s. The size, ecosystem requirements, and migratory ways of cranes make them vulnerable to a wide range of threats: loss and degradation of wetlands and grasslands, growing pressures on river systems and other freshwaters, illegal wildlife trade, poaching, poisoning, climate change. In many places, cranes have assumed the role of flagship species, catalysts for efforts to protect and restore biological diversity and ecological health in the ecosystems where they occur.[1]

The seasonal dialogue between wild cranes and their captive cousins at ICF is symbolic of the reality that Earth's biological diversity faces generally. As ecosystems of all types are progressively modified, their functions disrupted, and their constituent species and populations threatened, the captive breeding and reintroduction of endangered species has become more urgent and necessary. But such programs are expensive and labor intensive. They face major biological and ecological obstacles to success. They do not address the forces affecting the places where our fellow creatures evolved and in which they subsist. The conversation among the cranes thus echoes our own discussions about how to keep and restore a biologically diverse world, how to connect the wild and the captive, and how to foster a culture of care for human and natural communities.[2]

Despite the vulnerable status of many species and populations, cranes are in a sense more fortunate than many other endangered forms of wildlife. They are among the most intensively studied bird families in the world. In most of their home places around the world, they are valued, even revered. Their beauty, behaviors, and cultural resonance provide special opportunities for conservation education and action. Cranes also have the benefit of their own dedicated nonprofit organization. Since its founding in 1973, ICF has served as the focal point for an expanding global network of researchers, conservationists, and communities. Working through colleagues in more than a hundred countries around the world, ICF helps coordinate in situ field work in the cranes' native landscapes and ecosystems and the ex situ activities of laboratories, breeding and rearing facilities, and education centers.[3]

In this coordinating role ICF serves as a case study in the evolving relationship between zoos and the wild and provides an unusual model for zoos as centers for conservation education, action, and leadership. ICF is one of the 238 accredited members of AZA, the Association of Zoos and Aquariums. For many of its visitors, ICF is in essence a crane zoo. However, unlike most other AZA members, ICF is not a zoo that grew into a broader conservation mission. ICF began with conservation *as* its mission. A relatively small proportion of its budget is devoted to captive breeding and other on-site "zoo" activities. The proportion that goes directly to conservation work is the highest of any AZA member organization. From its beginning ICF has connected and integrated in situ and ex situ actions. In has always been, in effect, an *inter-situ* organization, linking its work within human institutions, communities, and landscapes to its work with wild birds, populations, and whole ecosystems.[4]

In its early years, ICF devoted special attention to developing captive breeding techniques and programs to ensure the survival of the rarer crane species. At the time, information on their status in the wild was often scarce and scattered. Captive breeding is now a relatively small part of a broad portfolio of conservation programs. ICF works around the world to safeguard crane populations and their ecosystems, watersheds, and flyways. It combines field programs with efforts to enhance local livelihoods and bring people together in community-based projects. It addresses larger-scale issues by securing water resources, championing land stewardship, conserving biodiversity on agricultural lands, and promoting clean energy and climate change solutions. It advances policy and develops conservation leadership. ICF's "zoo" functions continue, but in service to a mission that connects its visitors to the wide world of conservation.

The migrating sandhill cranes, calling out to their extended biological family at ICF, provide that connection for all who care. Only a few decades ago, the sandhill's voice was all but stilled in the upper Midwest. Reduced in the 1930s to only a few dozen wild birds persisting in the most remote wetlands of Wisconsin and Michigan,

FIGURE 2.1. Sandhill cranes coming into their evening roost.

Source: Photo by Ted Thousand. Courtesy of the International Crane Foundation.

sandhill cranes have since reclaimed their ancestral range across the region. Thanks to a combination of effective hunting regulation, education, research, and ecosystem restoration, Wisconsin alone now hosts more than twelve thousand resident cranes. Over the last five decades, the population's range has expanded across Wisconsin and into neighboring Minnesota, Iowa, Illinois, and Upper Michigan. As the high-flying sandhill cranes continue on their migratory way (figure 2.1), this is the encouraging message they can share with their family below.

The abbreviated word *zoo* first appeared in print in 1847, in reference to the London Zoological Gardens, known first as the "Gardens and Menagerie of the Zoological Society of London."[5] The London Zoo provided a Victorian template for the zoo as an institution. It existed as a walled compound, delimiting artificial space. It featured caged animals expropriated from distant lands, ostensibly for educational and scientific purposes. Behind the London Zoo lay precursors in royal menageries across medieval and Renaissance Europe. The modern zoo was thus imperial in its origins and colonial in its character. Zoo animals existed as objects—removed from their home landscapes, displayed in enclosed environments, deprived of the ecological and cultural relationships through which they evolved. Their agency as free-willed creatures was lost, confined behind bars, within the walls of the zoo, embedded in global empire, all restricted by a worldview of separation, control, and domination.

Yet the zoo is only the modern manifestation of a much older conception of the human place amid more-than-human space. The notion of spontaneous nature restricted to and controlled behind boundaries created for human benefit is ancient in language and myth. *Paradise* comes to us from the French, by way of Latin and Greek, via the Old Iranian for "walled enclosure." The word ultimately derives from a root meaning "to set up a wall."[6] The word carries us back, then, all the way to the Garden of Eden, to preconscious and prelapsarian hominids living amid kindred life, passing no judgment on the worth or value or beauty of the creation. Expelled from paradise, humans would then consciously bifurcate nature into domains of good and evil, sacred and profane, useful and useless, beautiful and base, crop and weed. In the dominant Western tradition, then, the place of the wild was wrapped up and warped in mythology from the get-go.[7] At the zoo, we hold the wild captive and find ourselves captivated. Here we go to reclaim a part of our humanity, and our world, that dimmed so long ago.

But is this original segregation of the walled and the wild endemic to the Western mindset? The early collection and confinement of wild animals has been documented among elites elsewhere—in ancient Egypt, India, China. Montezuma's zoo in Tenochtitlan was "an Aztec paradise filled with amazing animals, beautiful plants . . .

and even humans."[8] This suggests a more general, if not universal, human predilection for separation, identification, domination, and finally fetishization of the wild and exotic as civilizations developed, wealth accumulated, and social hierarchies emerged.[9]

Whatever the origins of the hardened binary between the cultural and natural, it now maps onto related dualisms that have defined debates and developments in conservation for more than a generation.[10] Scholars in geography, environmental history and ethics, anthropology, and related fields have been discoursing vigorously over the meaning of the wild and the humanized, and what it means for conservation.[11] Distilled for human consumption, these debates come down to an insistence that "traditional" conservationists have, since the movement's founding, been enthralled by a "myth of the pristine." The charge is that conservationists have focused narrowly (if not exclusively) on protecting an illusory wilderness that denies history and the agency of people in shaping landscapes, most especially Indigenous and local people. True and tragic enough, others reply. But to replace a myth of the pristine with a "myth of the humanized" is also suspect.[12] As pervasive and persistent as human impacts have long been, so has been the agency of lives beyond the human—call it wildness, or self-willed nature, or the sacred, or the kindred. Porous boundaries allow the wild and human to intersect, interact, and interpenetrate at every scale, from the gut to the globe.

Meanwhile, the "traditional" conservation movement is not so easily caricatured. Conservation (whether called that or not) has long demonstrated a commitment to lands and waters beyond the borders of protected areas, to nurturing ecological diversity and community resilience, and to active biocultural restoration work in the context of a complex and dynamic socioecological world.[13] Especially over the last generation, conservation has evolved away from the colonial "fortress" model to embrace a wide array of community-based approaches, mindful of social and ecological change, increasingly led by local, Indigenous, and place-based people and organizations.[14] Conservation exists, always and everywhere, in an *inter-situ* world, acting between the domestic and wild, amid the entire coevolving community of people, other animals, plants, fungi, and microbes, air and soil and rock and water, the oceans and the atmosphere.

The recent evolution of zoos has both reflected and partaken of these shifts in conservation. In 1994, George Rabb, the director of the Brookfield Zoo and a conservation biologist, described zoos moving through stages, "from menagerie and living museum" to "conservation centers in their communities."[15] In 1995, William Conway of the Wildlife Conservation Society wrote, "Wildlife communities . . . are not static. They change constantly in makeup and form."[16] In the last four decades Rabb, Conway, and other leaders in the zoo world began to envision zoos as places devoted explicitly to the well-being of whole ecosystems and the survival of all species. They urged zoos to involve themselves in holistic approaches to conservation through collaborative programs, immersing their visitors in dynamic exhibits that highlighted their status in the wild. This shift in the paradigm of zoos paralleled that which has been redefining conservation in general, away from top-down, coercive tactics and toward more inclusive community-based cooperative approaches.

Zoos and aquariums may have been established as segregated spaces, largely uninvolved and perhaps even uninterested in the living worlds where their inhabitants originated. But increasingly they are assuming a role at the nexus of the human and the wild, education and action, curiosity and commitment. No longer merely sites where animals are displayed as objects of entertainment and amusement, zoos have become nodes in networks of conservation that connect the remote wild and the locally human. That is to say, they exist increasingly as *inter-situ* institutions in a rapidly changing and relatively wild world.

What happens when we reimagine zoos not merely as *human* institutions comprising zookeepers, administrators, visitors, docents, and donors but as communities themselves, comprising humans, other animals, plants, waters and soils? Once we reject the binaries and depolarize the concept of zoos (and aquaria and botanical gardens) as segregated spaces, we may imagine new ways to integrate them into the larger socioecological systems in which they are invariably embedded and with which they continue to coevolve. We can begin to decolonize the zoo, deobjectify their inhabitants, and re-place ourselves as humans in our relations with them. We can allow the zoo to move more quickly away from its imperial and dominionistic origins to a relational future informed by an expansive ethic of care.[17]

Perhaps we will come to a time when our caring can lower the barriers between in situ and ex situ realities. Perhaps debates about the ideologies of inherent wildness and inevitable human cleverness will relax. Perhaps our conversations about humans and nature will be able to flow more freely in both directions. Until then, we peer through the porous boundaries with mixed feelings and longings, holding both the wonder and the irony. We echo the conversation between free-ranging cranes flying overhead and their confined kin below.

The fifteen species of cranes all inhabit wetland and grassland ecosystems but vary widely in their specific needs and their ability to adapt to human development, proximity, and activity. The sandhill crane, the most abundant of all cranes, is a generalist, occurring in ecosystems from Arctic tundra to temperate marshlands to tropical pine savannas. Once restricted mainly to the wildest wetlands, the species has recovered and even flourished in part because it proved to be behaviorally adaptable and tolerant of people (especially our agricultural fields). Sandhills now stalk the lawns of Florida backyards and Midwest college campuses. The Eurasian crane, the second-most abundant species, has in recent decades expanded back into portions of Western Europe where it has not bred for centuries.

By contrast, the more specialized species are less resilient, less adept at living among people and their encumbrances. This is especially true of the larger, whiter species that are more dependent on extensive intact wetlands: North America's whooping crane, Eurasia's Siberian cranes, East Asia's red-crowned crane, and Africa's wattled crane. The whooping crane exemplifies the contrast. Co-occurring historically in regions where the sandhill crane thrives, whoopers remain rare and wary, their population recovering steadily over the last seventy years, but much more slowly than the sandhills.

The Indian and Eastern subspecies of the sarus crane, the tallest flying bird in the world, present an exceptional case. Sarus cranes are rather specialized, yet these subspecies persist in some of the most densely populated areas of South Asia. They are able to do so because they face few direct threats from people in these portions of the species range. The cranes benefit from the profound cultural protection they receive in landscapes where Hinduism and Buddhism prevails. In a world of mingled human and wild realities, its story of coexistence may be the most important of all.

Such fine details of crane biology, ecology, history, and conservation are not easy to convey to the general public, but that has been the stock-in-trade of ICF's education programs since its inception. To provide greater opportunities to tell these stories and also to provide the captive cranes with more naturalistic settings, ICF embarked in 2018 on an extensive renovation of its visitors' area. Each of the renovated crane pens now features a constructed wetland and provides much more room to roam.

Flanking each pen is a panoramic mural inspired by the places where ICF works. The muralist Jay Jocham has portrayed the wetlands, grasslands, and savannas that the cranes use and the other animals and plants that live there. But each mural also depicts distinctive human elements (including the threats to cranes) of the cranes' native landscapes: farms and villages, pastures and livestock, power lines and fences, temples and cities.[18] The murals show explicitly that cranes live in these places alongside people, and vice versa. Cultures and creatures, the human and more-than-human, coexist and interact in the murals as they do in reality. The scenes capture both the perils and the promise of that coexistence.

One need only look through the nearby chain-link fence that marks ICF's own property boundary to know this reality and the real-world challenges and opportunities that boundary holds for crane conservation. ICF sits in a mixed landscape of farms, wetlands, upland grasslands, and forests. It is an optimal landscape for sandhill cranes and hosts one of the densest populations of breeding cranes on Earth. Meanwhile, just a mile down the road from ICF, an outlier population of three whooping cranes has taken up residence in the last few years, trying to find a toehold in this distinctly semiwild, semihuman agroecosystem. The in situ and ex situ roost here together, cheek by bill, in *inter-situ* space.

Another conversation has also been gaining momentum in this same landscape.[19] By a happenstance of history, the International Crane Foundation's campus is a direct neighbor to a community of the Ho-Chunk (Hoocąk) Nation, upon whose ancestral lands ICF sits. The U.S. government confiscated the tribe's lands across central and southwestern Wisconsin through a series of egregious treaties imposed in 1829, 1832, and 1837. At ICF and elsewhere, discussions between members of the Nation and descendants of Euro-American settlers have begun to provide long overdue acknowledgment of the historic trauma, loss, and resilience borne by the land and its people. This record includes the endangerment of, and more recent efforts to sustain and revive, the Ho-Chunk language. In recognition of this legacy and to begin work toward healing, ICF has consulted with the Nation and incorporated Ho-Chunk language in its new interpretive signs. Fittingly so. The tribe's name translates as "The

People of the Sacred Voice." Visitors may now learn the Ho-Chunk words for the birds with the great voices: the sandhill crane is peeją (*peh-jah*); the white whooping crane is péčąsą́na (*peh-cha-sah-nah*).[20]

In speaking and sharing these words, a necessary conversation can grow and begin to liberate cranes and people from the binary binds through which we confine ourselves. We may come to the zoo to relate to cranes and other creatures as "a communion of subjects, not a collection of objects" (invoking the words of the cultural historian Thomas Berry).[21] We may come fully to regard animals and plants not as "others" for humans to pluck from their home places to display, commodify, and consume but as fellow beings. As members with whom we are enmeshed in our shared communities. As kin.

The recent renovation at ICF has offered one more unexpected lesson. The work of remodeling ICF's facilities required closing the campus to visitors through 2019. Construction was completed in early 2020, but the emergence of the COVID-19 virus and the onset of the pandemic prevented ICF from reopening. The gates to ICF remained closed. The conversations among the cranes continued, but discussions within the global community of crane conservationists migrated over to Zoom. The coronavirus reminded us in the harshest way that the wild and the human comprise in fact one whole world. This liminal world of ours is subject, more than ever, to rapid and interconnected social and environmental change. In this *inter-situ* reality we pursue the work of biocultural restoration, healing people and nature together, and the ties that bind us all across time and space.

NOTES

The first portion of this chapter is adapted from Curt Meine, "The Calling of Cranes," *World Birdwatch* 15, no. 4 (December 1993): 14–17;

revised and republished by the Center for Humans and Nature (2016), https://www.humansandnature.org/the-calling-of-cranes; and as "Turnalarin Çağrisi," in *Anadolu Turnalari* (Anatolian cranes), ed. U. Ozdag (Ankara: Ürün Yayınları, 2019), 35–40.

I am grateful to all the participants in the Wilder Kingdom workshop and our hosts at the Arizona-Sonora Desert Museum (ASDM) for their stimulating input and exploration of ideas in this chapter. My colleagues at the International Crane Foundation have for decades provided vision and leadership in carrying out work at the boundary of the human and the wild. I am especially grateful to the ICF's cofounder George Archibald and president, Rich Beilfuss, for their assistance and input in developing the ideas here. Finally, I am deeply indebted to my late mentors George Rabb and Bill Conway and dedicate this chapter to their good memory. Rabb, who died in 2017, was a longtime supporter of the work of ICF (and of countless conservation biologists and organizations around the world). Bill Conway was instrumental in the founding of ICF in 1973 and remained involved in its work ever after. He passed away in October 2021, just three weeks before our "Wilder Kingdom" gathering.

1. For an overview of the conservation status and needs of the world's cranes, see George W. Archibald and Curt Meine, "Introduction to the Crane Conservation Strategy," in *Crane Conservation Strategy*, ed. Claire M. Mirande and James T. Harris (Baraboo, WI: IUCN Species Survival Commission, Crane Specialist Group, 2019), 9–16. See also Peter Matthiessen, *The Birds of Heaven: Travels with Cranes* (New York: Macmillan, 2001).
2. George Rabb and Kevin Ogorzalek, "Caring to Unify the Future of Conservation," *Minding Nature* 11, no. 1 (2018): 12–20, https://www.humansandnature.org/caring-to-unify-the-future-of-conservation.
3. See the website of the International Crane Foundation, http://www.savingcranes.org.
4. The Arizona-Sonora Desert Museum provides a comparable example of such a hybrid institution. For further consideration of the ex situ/in situ dualism and *inter-situ* connections, see Donald A. Falk, "Integrated Conservation Strategies for Endangered Plants," *Natural Areas Journal* 7 (1987): 118–23; David A. Burney and Lida Pigott Burney,

"Paleoecology and 'Inter-situ' Restoration on Kaua'i, Hawai'i," *Frontiers in Ecology and the Environment* 5, no. 9 (2007): 483–90; Kent H. Redford, Deborah B. Jensen, and James J. Breheny, "Integrating the Captive and the Wild," *Science* 338, no. 6111 (2012): 1157–58; and Irus Braverman, *Wild Life: The Institution of Nature* (Stanford, CA: Stanford University Press, 2015). Ben Minteer and James Collins similarly explore the need for "pan-situ" approaches in "Ecological Ethics in Captivity: Balancing Values and Responsibilities in Zoo and Aquarium Research Under Rapid Global Change," *ILAR Journal* 54 (2013): 41–51.

5. The Paris Zoo, widely regarded as the first modern zoo, was established in 1793 when leaders of the French Revolution relocated the menageries of French aristocrats to the Ménagerie du Jardin des Plantes. Michael James Graetz, "The Role of Architectural Design in Promoting the Social Objectives of Zoos: A Study of Zoo Exhibit Design with Reference to Selected Exhibits in Singapore Zoological Gardens," master's thesis, National University of Singapore, 1996. Graetz writes that "the first . . . animal collection kept for scientific purposes originated earlier (in 1624) as the royal Menagerie du Parc at Versailles" and that "Modern zoos are often said to have begun with London's Regent's Park Zoo in the 1820's because it was founded with a scientific purpose."
6. See https://en.wikipedia.org/wiki/Paradise.
7. Indeed, "the first recorded zoo containing animals such as lions is probably the royal menagerie of King Shilgai of the 3rd Dynasty of Ur, about 2000 BC, near Nippur in Mesopotamia"; Stephen St. Chad Bostock, "The Moral Justification for Keeping Animals in Captivity," PhD diss., University of Glasgow, 1987, 27. See also Stephen St. C. Bostock, *Zoos and Animal Rights: The Ethics of Keeping Animals* (London: Routledge, 2003). On the broader theme, see Nigel Calder, *Eden Was No Garden: An Inquiry Into the Environment of Man* (New York: Holt, Rinehart and Winston, 1967); Paul Shepard, *The Tender Carnivore and the Sacred Game* (New York: Charles Scribner's Sons, 1973); Evan Eisenberg, *The Ecology of Eden: An Inquiry Into the Dream of Paradise and a New Vision of Our Role in Nature* (New York: Knopf, 1998); and Carolyn Merchant, *Reinventing Eden: The Fate of Nature in Western Culture* (New York: Routledge, 2004).
8. Natalia Klimczak, "Montezuma's Zoo: A Legendary Treasure of the Aztec Empire," March 18, 2020, https://www.ancient-origins.net

/ancient-places-americas/montezuma-zoo-legendary-treasure-aztec-empire-005090.

9. For a sweeping argument for the progressive humanization and domestication of the global environment, see Erle C. Ellis, "Ecology in an Anthropogenic Biosphere," *Ecological Monographs* 85, no. 3 (2015): 287–331, https://doi.org/10.1890/14-2274.1. For counterarguments, see George Wuerthner, Eileen Crist, and Tom Butler, eds., *Keeping the Wild: Against the Domestication of Earth* (Washington, DC: Island, 2014). See also Fatih Uenal et al., "The Roots of Ecological Dominance Orientation: Assessing Individual Preferences for an Anthropocentric and Hierarchically Organized World," *Journal of Environmental Psychology* (2022), 101783.
10. For a discussion of dualisms in ecological restoration, see Jozef Keulartz, "Boundary Work in Ecological Restoration," *Environmental Philosophy* 6, no. 1 (2009): 35–55.
11. For compilations of relevant scholarship and commentary, see J. Baird Callicott and Michael P. Nelson, eds., *The Great New Wilderness Debate* (Athens: University of Georgia Press, 1998); Michael P. Nelson and J. Baird Callicott, eds., *The Wilderness Debate Rages On: Continuing the Great New Wilderness Debate* (Athens: University of Georgia Press, 2008); Ben A. Minteer and Stephen J. Pyne, eds., *After Preservation: Saving American Nature in the Age of Humans* (Chicago: University of Chicago Press, 2015); and Gavin Van Horn and John Hausdoerffer, eds., *Wildness: Relations of People and Place* (Chicago: University of Chicago Press, 2017).
12. See Thomas Vale, ed., *Fire, Native Peoples, and the Natural Landscape* (Washington, DC: Island, 2013); and Wuerthner et al., *Keeping the Wild*.
13. See Gary Paul Nabhan, *Cultures of Habitat: On Nature, Culture, and Story* (Washington, DC: Counterpoint, 1997); Curt Meine, "A Letter to the Editors: In Defense of the Relative Wild," in *After Preservation*, 84–95; Paddy Woodworth, *Our Once and Future Planet: Restoring the World in the Climate Change Century* (Chicago: University of Chicago Press, 2013); Curt Meine, "Restoration and 'Novel Ecosystems': Priority or Paradox?," *Annals of the Missouri Botanical Garden* 102, no. 2 (2017): 217–26.
14. The literature on community-based conservation is extensive. A helpful launching point for exploration is Fikret Berkes, "Community-Based Conservation in a Globalized World," *Proceedings of the National Academy of Sciences* 104, no. 39 (2007): 15188–93. For foundational works

in traditional ecological knowledge, see Dennis Martinez, "Managing a Precarious Balance: Wilderness Versus Sustainable Forestry," *Winds of Change* 8, no. 3 (1993): 23–28; Enrique Salmón, "Kincentric Ecology: Indigenous Perceptions of the Human–Nature Relationship," *Ecological Applications* 10, no. 5 (2000): 1327–32; Robin Wall Kimmerer, "Weaving Traditional Ecological Knowledge into Biological Education: A Call to Action," *BioScience* 52, no. 5 (2002): 432–38.

15. George B. Rabb, "The Changing Roles of Zoological Parks in Conserving Biological Diversity," *American Zoologist* 34, no. 1 (1994): 162. See also George B. Rabb, "The Evolution of Zoos from Menageries to Centers of Conservation and Caring," *Curator: The Museum Journal* 47, no. 3 (2004): 237–46.
16. William Conway, "Zoo Conservation and Ethical Paradoxes," in *Ethics on the Ark: Zoos, Animal Welfare, and Wildlife Conservation*, ed. Bryan G. Norton, Michael Hutchins, Terry Maple, and Elizabeth Stevens (Washington, DC: Smithsonian Institution, 2012), 1–9.
17. Rabb and Ogorzalek, "Caring to Unify the Future of Conservation." Bostock, in *Zoos and Animal Rights*, envisions the zoo evolving to become "an acceptable community of animals" (182). See also George B. Rabb and Carol D. Saunders, "The Future of Zoos and Aquariums: Conservation and Caring," *International Zoo Yearbook* 39, no. 1 (2005): 1–26.
18. See "Behind-the-Scenes Look at Painting the International Crane Foundation Exhibit Murals," YouTube video, https://www.youtube.com/watch?v=IT1TPEDc2Xw.
19. See Curt Meine, "Healing Sacred Earth," in *Kinship: Belonging in a World of Relations*, vol. 2: *Place*, ed. Gavin Van Horn, Robin Wall Kimmerer, and John Hausdoerffer (Libertyville, IL: Center for Humans and Nature Press, 2021), 126–35.
20. Hočąk Lexicon, https://hotcakencyclopedia.com/ho.HocakLexicon.html.
21. Thomas Berry, *Evening Thoughts: Reflecting on Earth as a Sacred Community* (Berkeley, CA: Counterpoint, 2015), 17. See Paul Waldau and Kimberley Patton, ed., *A Communion of Subjects: Animals in Religion, Science, and Ethics* (New York: Columbia University Press, 2006).

3

ANIMAL ART AND THE CHANGING MEANINGS OF THE WILD

ALISON HAWTHORNE DEMING

I built a trail on my family's property on Grand Manan Island in the Canadian Maritimes a decade ago. It covered steep ground over basaltic cobble and up into black spruce, ash, and balsam fir, then into a meadow of lacey ferns that had taken advantage of the disturbance of a logging operation. The trail rises up three steep pitches, then levels through an area I call the moss rocks and another I call the sapling narrows, intersects an old logging road overgrown with bracken fern, cuts into a thicket where I hear hermit thrushes trill though I never see them, and on to ground softened by peat and the primeval bog, the wildest part of the trail, and the place that makes children scream and giggle in delight as their sneakers sink into mud. There's a vernal pool where deer and raccoon drink. I've seen their tracks though rarely see them. The trail covers a couple of miles. The farther into the wild it goes, the messier it gets with deadfalls and brush. It's rough going but thrilling in what it represents for me. The trail marks the land's response to my desire to know it better than I did for the decades when this forest remained an abstract notion to me. It marks a trans-species getting-to-know-you conversation, and I swear on one of my forays I heard a stand of white birches call out to me and ask for my attention. Why was I paying attention to those ferns when *they* were standing statuesque and beautiful at the meadow's edge? OK, OK, I replied, and I have never again ignored them.

The pandemic meant that I could not visit the island for two summers, Canada being perspicacious about the risks attached to invasive critters south of their border. By the time I returned, the trail had gone feral. It's amazing how much a black spruce sapling can grow in two summers. The wild had taken over enough to obscure the trail and redefine the topography I thought I knew by heart. But with clippers and Weedwacker and a feisty woman pal who's masterful with a chainsaw, we dispatched enough of the deadfalls and overgrowth for the trail to once again become a reality. The sense of the wild that inheres in this story suggests uncontrolled growth. In that sense, *Homo sapiens* is the wildest creature of all.

When the word "wild" comes up in casual conversation these days, it often speaks of an irreverence to social norms. Those kids had "a wild week" on spring break. We watch old *SNL* clips of "two wild and crazy guys." Insurrectionists bent on disrupting the peaceful transfer of democratic power erupt into what they called "a wild protest"—an action opposite to what Thoreau or Gandhi or King meant by civil disobedience. Wild is unruly, disruptive, insubordinate, violent, destructive, cruel. The iconic figure of the January 6, 2021, uprising wears a headdress graced with bison horns, face draped with raccoon tails that reach mid-chest. He lifts his head and howls like a wolf to complete his invocation of animal power, though his actions have nothing to do with animal prowess and everything to do with the human capacity for self-delusion.

What's missing from such vernacular usage of the word "wild" is reference to how the wild world works. Wild is inventive, adaptive, relational, communitarian, committed to survival. Wild is beautiful. Wild is nurturing. Wild is fierce. Wild brings formal order to chaos. Wild organizes itself into species and habitats. Wild makes do with what is given. Wild creates fertile ground for what comes after. Wild in the Anthropocene is diminished. Wild is constrained by human actions. Wild is held captive in shrinking islands of habitat and the climate weirding caused by our carbonation of the atmosphere. Our bodies are wild assemblages of microorganisms. Our minds are wild seas filled with wild emergent properties. Yet wild now is contingent on human dominion. How can it be that we are driving the animals from their homes? Just check out anyone's Facebook or Twitter account to see how much we love animals. The cognitive dissonance between our affections and our actions is staggering. Is "wild" even a thing anymore on our zoo-planet, the animal world caged by the anthropogenic constraints of habitat loss and climate change?

I was asked to speculate about how zoos and wildlife parks might be reimagined in the Anthropocene. What would a "wilder zoo" look like? What on Earth can it mean to speak of creating a "wilder zoo," when we know that everything "wild" is subject to human dominion? Can zoos help turn our affection for animals into ethical regard and action that will preserve space for them, a future for them, on our ravaged planet?

A good place to start answering that question is with another question: what are zoos for? Entertainment, education, conservation, sanctuary. We go to the zoo to see the animals and to see ourselves more clearly in the mirror and mystery of their presence. We go to the zoo to learn of the diminishment of the animal world. We go to the zoo to forget the diminishment of the animal world and be swept away in wonder with being in the animal presence. People have collected animals for study, observation, and pleasure for over five thousand years—and probably well beyond. Excavations in Egypt turned up remains of a menagerie dating to 3500 BCE hosting hippos, elephants, baboons, and more. Ancient Chinese royalty of the second century BCE built a "House of Deer" and a "Garden of Intelligence." But zoos became "zoos" when the London Zoological gardens opened in 1828 for zoological studies. Once people went to zoos to learn about creatures that populated a world full of marvels in places few would have the resources so visit. Zoos expanded our imaginations to the world's beauty and bounty.

Now zoos give visitors a "last chance to see" (to borrow Douglas Adams's turn of phrase) some of the magnificent beings that are being driven from Earth by the witless project of mass ecocide. Zoos now expand our imaginations to the animal world's diminishment and collective grief—a trans-species grief, if we are to take seriously the emotional capacities of our animal compatriots. Zoos are meant to foster education and empathy, nourishing head and heart. Zoos have prioritized science education and natural history as ways to know animals, but taking the long view of history, I believe that art—literary, visual, musical—has played at least as important a role in leading head and heart toward deeper engagement with the animal world. And a synthesis of aesthetics and science, a tradition kick-started nearly two centuries ago by the works of Alexander von Humboldt, may just provide the binocular view for seeing the world anew and fostering change. Something transformative happens when we reach across disciplinary divides to understand one another's language.

In my book *Zoologies* I explored the question of what animals mean to the human imagination and how art contributes to that meaning making. In the introduction, I wrote:

> Animals surrounded our ancestors. Animals were their food, clothes, adversaries, companions, jokes, and their gods. In this age of mass extinction and the industrialization of life, it is hard to touch the skin of this long and deep companionship . . .
>
> What do animals mean to the contemporary imagination? We do not know. Or we have forgotten. Or we are too busy to notice.

Or we experience psychic numbing to cope with the scale of extinctions and we feel nothing.

I began my noticing with art made thirty thousand years ago, the cave paintings of Lascaux and Chauvet, and even earlier thumb-sized animal figurines. A lion-man carved from mammoth ivory was unearthed from a cave in the Swabian Alps. The thumb-sized chimera dates from about 42,000 years ago. It is polished smooth, as if from rubbing, as one would worry a stone kept in the pocket. The craftsmanship is skillful, the lion standing upright with a finely detailed head. The object would have required many hours of carving, an investment of time that speaks of its value to the maker. The lion figure telegraphs the deep human urge to be close to—unified with—the power and mystery of wild animals. This is the prehistoric period during which caves in southern Europe were being transformed into galleries lined with animal images: bison, horse, aurochs, deer, rhinoceros, and many more. The drawings were made on cave walls deep in the earth, drawn by torchlight in the stony dark. The images had to be held in mind while the artist ventured into an underground chamber far from the realm where observations would have occurred. The wild gives rise to the aesthetic imagination.

The drawings represent accurate natural history observation. In the Chauvet cave, four horses line up neck to neck as if a running herd, heads slightly bowed, as horses do when running. The details are beautifully rendered in terms of phenology, and they reflect the behavior of horses. Their manes are brushy and short, like those of Przewalski horses, the last wild subspecies of horse. These art works may represent a deep sense of belonging in the animal world. Or they may represent the longing that it might be so. Some thinkers believe the origin of art lies in a dawning awareness of human separation from the animal world and the desire to restore the sense of belonging through acts of creation. Bearing witness to the animal world required more than simply seeing; it required record keeping that was both accurate and aesthetically transformative. Are such drawings art or science? At this time in the human past, there was no distinction between these modes of seeing. The accuracy speaks of science—a record of field observation. The beauty speaks of art. And the human handprint blown with pigment onto the cave wall speaks of art's longstanding need to bear the mark of self, the urge to participate in acts of creation: I am the maker.

Much of the art that we see in zoos is illustrative. As with interpretive texts, it seeks to inform and educate with accurate depiction of animals and their habitats. It supports the general feeling of visual spectacle that can be the zoo experience. It communicates the majesty of wild animals and offers a view so detailed it makes the viewer lust for such close encounters with the wild. The wildlife paintings by the German artist Carl Rungius of the 1920s and 1930s are exemplars. He believed he could capture the colors of nature only if he painted outdoors. The plush red fox posing on an alpine boulder. The moose pair depicted *During the Rut* in their swampy domain, bull in laconic pursuit of cow—the moment held just before the action begins. We can feel the latent sexual tension drawing us in commonality with the creature and his desire, despite his outsize otherness. These are idealized animals, noble animals, wild animals held as archetypes for their species. Their habitats are depicted as gorgeous landscapes, no interwoven web of critters to keep these celebrities company or challenge their dominion. They stir in us that rise in spirit that beauty generates, that sigh of recognition at nature's marvels. Is that empathy? I'm not sure. Sometimes I think this sort of image creates more empathy for ourselves. If only we were so free and magnificent and independent. For all of our brilliance, we seem diminished creatures in juxtaposition to these beauties.

The photographer Chris Jordan slams us into the twenty-first century with his body of work addressing the burden that mass consumerism and industrial growth have put on the natural world. Among his works is the film *Albatross* and photographs taken on Midway Atoll, a remote cluster of islands more than two thousand miles from the nearest continent, "where," he writes, "the detritus of our mass consumption surfaces in an astonishing place: inside the stomachs of thousands of dead baby albatrosses." His photos show the dead birds, bellies exposed full of elastic bands, bottle caps, disposable lighters, and unidentifiable plastic fragments. The albatross parents, foraging over the Pacific gyre, mistake the floating trash for food and feed lethal meals to their chicks. Jordan makes the easy analogy to our own behavior, in which the human appetite for stuff poisons us and the biosphere. But Jordan's intent is on healing through art. He writes: "What if facing the dark realities of our world, instead of being an experience of pain, anxiety, overwhelm, and hopelessness, could be a doorway into a place of beauty, connectedness, and healing?" The stirring of empathy these images arouse is a call to action. Jordan believes if we feel these things more deeply, they matter to us more and help us "to face the big question, which is: how do we change?"

I'd like to mention two additional artists, both of whom have had work exhibited in the Nevada Museum of Art's Triennial Art + Environment Conference, always offering a stunning array of international works that challenge and delight. First, Kate Clark, whose work brings us back to the chimera as an artistic form that transcends time and place from the Paleolithic to the Anthropocene. Clark works with taxidermied animals, as do many contemporary artists, from Robert Rauschenberg's goat to Damien Hirst's tiger shark, from Lola Stern's disturbing goat fashioned as a laundry basket to the Tlingit artist Nicholas Galanin's stunning *Inert Wolf* rearing to life from his flattened skin spread on the floor for a rug.

FIGURE 3.1. An exhibition of Kate Clark's work, *Licking the Plate* (2014), at the Nevada Museum of Art.

Source: Photo by the author.

Kate Clark's work uses animal skins—gazelle, zebra, kudu. She takes away the animal's natural facial features and creates a human face on the body of the animal. The result is arresting—a compelling chimera of woman and kudu—a contemplative human gazing out at the viewer from an animal body, ears and eyes alert (figure 3.1). Clark speaks of the language barrier that separates humans and animals and of her hope that by giving the animal a human face she makes possible communication across the divide. The work is "about a necessary balance between man and animal and not about the hierarchy of man over animal." Clark's work inspires for me the question: if animals had human faces, would we treat them with less cruelty? Perhaps, though we have plenty of evidence that the human capacity for cruelty is vast. However, I can't help but feel the ameliorating effect of those chimeric beings, the way that the images make it much harder to see the animal as "other." They stare us down. Do they mean to accuse or evoke our love and care? In either case, the chimera is a durable trope that inhabits the human imagination (don't get me started on goats spliced with spiders so that they produce silk rather than milk—a real thing) and offers fruitful ground for thought experiments and art projects that enhance the sense of our animality and the personhood of animals.

Cannupa Hanska Luger, a Native artist from Standing Rock, is of Mandan-Hidatsa-Arikara-Lakota-European dissent. He works in ceramics, textiles, installation, and performance art. His work is grounded in what he describes as his different perspective on environmental art, coming from "our dislocation from the land." He works to articulate that we *are* the land. We *are* the water. He works to counter the separation between people and place that was inflicted on the American West by colonizer and settler culture, as articulated by Kelsey Dayle John and Mariah ShieldChief in this collection.

Luger's work addresses what it means to belong to the land versus living a life devoted to belongings. He speaks of "indigenous futurism" and the "applied science embedded in indigenous cosmology." He speaks of buffalo as "a nation" to his people and the importance of this animal to his ancestors. Among the animal representations in his work is a ceramic buffalo skeleton that he has constructed, as if to both rebuild the animal and memorialize its death, in galleries and outdoors in streams where its "blood" (as red plastic strips) enters the water's flow. Much of his work is a collective act of creation, as embodied in his giant *She-Wolf* sculpture, made with a steel armature, ceramic eyes, and a repurposed fabric pelt adorned with over eight hundred embroidered bandanas sourced from communities across the world. Here is art that acknowledges the impact on human beings of species disappearing, art that calls the viewer to rebuild connections with the animal world and with our long human story.

I imagined that I might write an entire essay on the animal cartoons that appear in the *New Yorker*. I have been collecting them for years. I am fascinated that this most cosmopolitan publication is so obsessed with animals as entertainment and commentary on what it means to be human. And what it means to be animal in this most human age. A bespectacled gazelle is seated in an office chair, hoof pressing the buttons of his desk phone. "Miss Winter," he says. "Will you see if the lion has left the water cooler yet?" OK, it's dated in both phone tech and gender dynamics, but we have all known a lion in the workplace that we'd rather avoid. A wolf pack stares befuddled at a grimacing she-wolf who wears a transponder with antenna strapped to her back. The alpha says, "We found her wandering at the edge of the forest. She was raised by scientists." A couple of goldfish swim in a fish tank decked with fake castle and fake seaweed, enormous human face pressed to the glass. The fish wear anxious expressions. One tells the other, "I think it loves us." Grizzlies are catching airborne salmon in a river, while in the foreground two grizzlies adapted to our ways are playing a pinball machine on which digital salmon leap up the screen. The cartoon is a brilliant vehicle for education and empathy, for instructing us about how the animal world might perceive us and how we are merely animals despite all of our pretense.

I've privileged visual art in this essay in part because zoos are so much about visual spectacle. As a poet and essayist, I have had the opportunity to work on several projects bringing poetry and storytelling into play in the task of enhancing scientific literacy and empathy for the animal world. As poet-in-residence at the Jacksonville Zoo and Gardens, I headed a project to install a poetry path and accompanying community programs. I got to know zoo professionals—educators and keepers—and learn about Species Protection Plans that guide efforts to bring genetic diversity back to imperiled animals. But I cannot honestly say that I got to know any animals. That is, until I visited the Little Rock Zoo to tour their poetry installation. My host asked me if I'd like to meet "the girls." Sure, I said, bemused, not knowing what to expect. We entered an enclosure and met the mellow, white-haired keeper. He called two Asian elephants over to visit us. (Yes, I know, you may be asking what are Asian elephants doing in confinement in Little Rock, Arkansas, but that is a conversation for another day.) The elephants had toys to play with—bowling pins and a beer keg to toss around. They had a huge black ball to roll around—repurposed from an operation that blows the balls through an oil pipeline to clean it. The keeper told me that the elephants are so smart it's hard to keep them interested. He hides fruit for them to find, builds a cache up high so they have to work for food. He's always looking for new ways to keep them engaged.

These two girls were friendly, sidling up to us and idling while we talked. I stroked one's forehead while she leaned close, her elegant eyelashes, black and shiny, long as my thumb. I noticed a thick scar on her rear ankle. What's her story? I asked. When she arrived, the keeper explained, they had found a packing slip in her crate. She'd been a circus elephant shipped to New York City in 1951. She must have been chained by that ankle in the crate when she wasn't performing. She was sixty-one years old. When she got to Little Rock, she was afraid of everything: the peacocks, the kids' zoo train. The other elephant had been purchased from a logging camp in India where she had been employed. She was a few years younger. Their whole lives had been spent in human company, and the zoo was the kindest chapter. Suddenly these were not archetypal "Elephants" standing in for their species. They were individuals with life stories. Suddenly the zoo looked less like a prison and more like a sanctuary. Suddenly I was moved beyond wonder to love.

In a second project, I worked with the Milwaukee Public Library and Milwaukee Public Museum to curate community programming and a sidewalk poetry installation in which signage matched poem excerpts with a relevant science message and visual image (figure 3.2). Our theme was: how do you see yourself in the natural world? Each sign represented dual ways of knowing, bringing science and aesthetics together. Once we had a range of poems selected, the scientists at the museum worked to finalize the ones they could effectively match with content from the museum. Milwaukee's poet laureate Karla Huston's "Sea Change" was paired with an image showing both the current landscape of Wisconsin and that landscape as a seascape 430 million years ago. The design of the installation placed the art and science on equal footing, together serving the project goal of enhancing STEM learning through interdisciplinary collaboration.

What might we mean today by creating "a wilder zoo"? Our project venue, the Arizona-Sonora Desert Museum, isn't seen as a zoo. That's because though it exhibits animals and they are confined, it has a large enough footprint in the very habitat it celebrates that it

FIGURE 3.2. Sign from the Poetry Path linking Milwaukee Public Library and Milwaukee Public Museum.

Source: Image provided by the author.

feels at home in its place. And it features animals and plants of this place, taking important steps toward showing wildlife not as exotic celebrities but as relational members of a biologic community. "Wilder" is coming to mean "more relational" as we learn more and more about connectivities in the plant and animal world. Forests humming with rhizome networks, trees sharing resources with other trees, flowers communicating with bees by way of scent. We live within a worldwide web that has nothing to do with cyberspace. Framing interpretive materials to foreground biologic community is a good start at redefining "wild" to more accurately reflect this understanding.

A wilder zoo can speak truth to grief in poems, prayers, letters, or songs. A wilder zoo can acknowledge the full range of emotion that animals inspire in us. In the face of mass extinctions, grief may be as important to acknowledge in understanding the fate of the animal world as is the sense of wonder. Art is very good at communicating complex emotions. In the age of mass extinctions, grief and love run side by side. Grief is a measure of our love. I am particularly interested in how storytelling, as Gary Paul Nabhan has championed, can build a caring relationship between people and animals. Ask anyone to tell you an animal story, and with minimal prodding they will deliver. Knowing the individual life stories of animals we meet in a zoo can help us understand just what "the wild" means today rather than presenting an archaic notion of the wild as something untouched by human hands. Nothing is untouched by human hands these days, when our influence affects the air all creatures breathe and the water all creatures drink, when we are robbing creatures of their habitat at a breathtaking pace.

Empathy is a human capacity but not a given. The arts play a major role in educating our empathy—and understanding the inner wildness that can lead our actions to be destructively at odds with our values. That inner wildness—"wild mind," as Gary Snyder named it—is also the source of our brilliant human inventiveness. Educating our empathy has to do with aiming that wild mind toward more loving outcomes, a sentiment embodied in these lines from Wordsworth's "Prelude," which had a prominent placement in the Jacksonville Zoo and Gardens poetry path:

> What we have loved,
> Others will love,
> and we will teach them how.

We are on a steep learning curve in understanding what it means to be wild today. I'm all for having fun and creating thought experiments at the zoo, activities that move visitors from being consumers to being actors (and maybe this requires creating a very cool maker space): getting kids to dress up like a honeybees and do a waggle dance; having a scavenger hunt where people sleuth out, say, all the creatures they saw with wings, with fur, with scales, with fangs; having a story hour in which visitors can tell animal stories to one another—kids go first; having a quiet place where people can sit in contemplation and write a letter to an animal, or make a drawing of a chimera they create out of their favorite critters, or write a story about what the world would look like in one hundred years if their dreams came true about the fate of animals, or drawing a new map that reflects the big vision of designating half the Earth for the other-than-human animals.

I'll close with an anecdote that strikes me as instructive in terms of the complexities of what we mean when we contemplate "the wild" these days. I recently visited an online workshop for writers wanting to write on behalf of animals. One writer's question really hit me. She lived in Seattle. Coyotes had been showing up in her neighborhood and preying on domestic cats that people let out at night to hunt. The neighbors were at odds. The cat people wanted to extirpate the coyotes, but others were thrilled to see wild critters. Who wins? For me, it was an easy call. I've been a cat owner in the past, and one who let her cats roam freely—one who celebrated their prowess and took delight in seeing their delight at being wild and free. I thought it was cruel—against their nature—to keep them indoors. But after I learned of the devastating toll our furry feline darlings take on songbirds, I had a conversion experience. Our domestic cats are not wild—otherwise they would be naturally culled by predators. There are too many of us and too many of them. So whatever balance might once have existed in this predator/prey dance, that time has passed. This seems a simple task of education. Share the data about songbird deaths with the neighbors: a few billion songbirds killed by domestic cats each year in the United States alone. Then keep the damn cats indoors. But this asks for the cat owners to do both emotional and ethical work. Data alone does not make people change. People love their cats and respect their "wild" nature. Would acknowledging what the cat people are being asked to give up help? Would asking for stories about who their cats are and what matters to them help? A hard sell, to be sure. As for the coyotes, don't we want a world where wild animals are free to roam in our neighborhoods? Maybe and maybe not. We love the wild but want it boundaried. Just how much of our own desires are we willing to compromise for the sake of animals? We are on a steep learning curve, along which the questions will continue to come hard and fast. We will need every tool and salve in our kit.

4

CAN ZOOS CONNECT PEOPLE WITH WILDNESS?

SUSAN CLAYTON

Zoos have multiple functions, and these seem to have expanded over time. Many or most zoos focus on promoting positive outcomes for animal species and are directly involved in species conservation; they refer to themselves, with justification, as conservation organizations. This role is crucial. However, the original and still central purpose of the zoo was oriented toward outcomes for visitors rather than for nonhuman animals or species: displaying animals to be appreciated by a human audience. The anticipated impacts of such displays have traditionally included both entertainment and education, but more recently many zoos have focused on a third impact: connection. Zoo visitors can and should expect not only to have a nice outing and to learn something about animals but also to have an emotionally significant experience that makes them feel more connected to animals. These aims are reflected in the mission statements of many zoos;[1] the in-house magazine of the American Association of Zoos and Aquariums is even called *Connect.*

These diverse goals are important to understand because they can, or at least should, guide decisions about zoo design and programming. Such decisions are complex because the goals may be in competition: at a minimum, for example, allocating more resources of money or staff to entertainment may mean that there are fewer resources for conservation. More directly, the goals may impede one another. The move toward connection as a goal reflects the fact that there is only a weak link between the zoos' other aims: education has a disappointingly small relationship with conservation support, and entertainment, by objectifying animals, may in some cases even reduce conservation understanding and concern.[2] But encouraging people to feel a sense of connection to nature can help fill multiple goals. It is not only enjoyable for them but may increase support for conservation initiatives. Surveys of zoo visitors show that experiencing a sense of connection to an animal during a zoo visit is associated with interest in conservation for that animal or species.[3] But how does connection to an animal translate into support for a species or for environmental conservation more generally? It may lie in the perception that the animal embodies the wild.

THE SENSE OF CONNECTION

The goal of creating a connection to nature is complex and ambitious. Not every zoo visitor will leave feeling closer to animals or to the natural world, but research suggests that this is a potential outcome of a zoo visit.[4] What does it mean to talk about connection to nature? The concept has been described with various terms, including "connectedness to nature," "nature relatedness," "inclusion of nature in the self," and "environmental identity"; although these concepts are not identical, they share an emphasis on the feeling of connection. In general, a connection to nature can be said to include both an emotional and a perceptual response: a feeling of warmth or affection and a sense that nature is less distant, more related to the self. On the Arizona-Sonora Desert Museum website, the emotional and perceptual impacts are summarized as "love and understanding" of the Sonoran Desert.

When it comes to animals more specifically, one study asked zoo visitors to name an animal they connected with and describe what it meant to connect with that animal.[5] In open-ended responses, the visitors described themes of appreciation of the animal, attribution of mental states to the animal, emotions that the animal inspired in the visitor, interaction with the animal, and proximity to the animal. Some of these themes can be seen as directly reflecting the emotional part of connection (appreciation and inspiring emotions). The others relate to the perception of closeness or similarity: interaction with the animal or proximity to the animal suggests closeness, while attribution of mental states to the animal suggests a perception of similarity. Several visitors specifically mentioned a feeling of empathy associated with the animal.

Feeling a connection to nature reflects a perception not only of nature but of one's own relationship with the natural world. People who are connected to nature recognize their own interdependence, with a greater sense of the ways in which their own well-being is bound up with the well-being of the natural world. They want to spend more time in nature and value nature more. They give greater moral standing to natural entities, for example, supporting animal rights.[6] Just as a perception of similarity and empathy tends to promote helping, even among humans, people who perceive themselves as connected to nature are consistently higher in environmental concern and intention to engage in environmentally protective behavior.[7]

VALUING WILDNESS

Although people want to feel a connection to nature and to the animals in the zoo and may enjoy recognizing some similarities between themselves and the animal,[8] the zoo is also very importantly a place of difference. People go to the zoo to see what they don't see in their daily lives: the exotic, the large, the rare, even the endangered.[9] Even though zoo animals are clearly not living in the wild (nor, typically, were they born in the wild, though many zoo visitors do not realize this), they are still valued in some sense as "wild" animals, giving them an appeal that is distinct from the opportunity to connect with the animals that can be encountered in or near one's home. It can be interesting to observe young children, who have not yet learned the difference between familiar and unfamiliar, in a zoo; they can be just as excited to see a squirrel or chipmunk outside an exhibit as to observe the exotic animal inside an enclosure.

It is an odd thing about humans: despite the time and effort we spend transforming the environment to make it more comfortable or productive for us, we still like the idea of "wildness." In evaluating landscapes, people tend to give the highest ratings to those that show little evidence of human intervention and cite "wildness" as an important attribute. Studies that have examined people's reflected preferences show that many people do assign value to wilderness and are even willing to pay for it.[10] One reason for this may be an unconscious anticipation of possible human benefit. Wild nature has the potential to reveal unexplored and unexpected resources, and although most of us no longer directly obtain those resources for ourselves, people may intuitively respond to this aspect without realizing it. It is also true that there are cultural differences in attitudes toward wildness, which change according to the social context. (See, for example, the discussion of "wild" in the chapter by John and ShieldChief, which identifies some of the stereotypes and cultural context that surround the term.) It may be that people value "wilderness" in part when they no longer interact with nature as part of their daily lives.

Many people also value the opportunity to feel a connection to wild animals, such as whales or wolves. Perhaps such experiences allow people to feel accepted by the world as a whole—to feel that they belong in it, despite their reliance on modern technology. During the social disruptions caused by COVID-19, many people found hope in stories (not always accurate) about nature reclaiming urban places. Attitudes toward wildness may also reflect an attribute of an underlying biophilia—E. O. Wilson's term for the instinctive interest in and emotional response to the natural world.[11] Or perhaps wilderness simply represents freedom: a lack of rules and of social monitoring that is appealing in the abstract despite its shortcomings in practice.[12] Or all of these things.

Wild nature is also often described in terms of its emotional impact: people talk about it as a source of "transformative" experiences. People cite awe in response to the vastness or innumeracy of nature. When nature inspires awe, it may also promote humility and a recognition that a human-centered perspective is not the only one. In other words, transformative experiences reshape something fundamental—an identity, or value, or way of interacting with the world. Bryan Norton, similarly, thought that experiences of nature could promote transformative values, or values that were consistent with the ideas that humans and nature are interdependent.[13] Awe-inspiring nature may even bring out the best in people; some studies have shown that it can encourage ethical behavior and a sense of meaning.[14]

But people's reactions to wildness also have limitations. One is that people misjudge it—what they think is wild or a healthy ecosystem may be neither.[15] Another is that people typically diminish it; human activities endanger the health of wild nature. A third, from a practical point of view, is that when wilderness is perceived as separate from the places where people live, it is by definition difficult for people to get to, requiring some investment of resources (often including fossil fuels) to reach. And a final limitation is that, as the population of humans continues to explode, there is less and less room for wild nature to exist. The argument has been made that people should not even be encouraged to connect to wild nature because they will then have to cope with losing it.[16] Why (in a largely, but not completely, rhetorical question) form an emotional attachment that will just lead to grief?

ENCOUNTERING THE WILD AT THE ZOO

In such a context, perhaps zoos can help provide people with the access to "wildness" that they want and serve to promote connections to a healthy environment without sending them out into a fragile and disappearing wilderness. Zoo exhibits already satisfy some of what we know about restorative environments.[17] They are fascinating; that is, they effortlessly attract interest. They reflect an environment that is separate from the visitors' day-to-day life. Exhibit designers are experts at creating the illusion of extent and depth even with limited space. But restorative environments also need to be compatible with visitor goals. Visitors come to be entertained by seeing something new and different, to occupy their children, and perhaps to learn something, so exhibits should be designed with these goals in mind.

Exhibits could also focus on meeting the goal of connection, by highlighting relevant aspects of identity. Providing specific and individuating information about an animal or drawing attention to similarities between the animals and the visitor may encourage visitors to feel a sense of shared identity with the animal.[18] In addition, many zoos are interested in helping visitors learn about local species and landscapes. Emphasizing the shared place identity and local cultural heritage may also help establish a sense of connection.

To embody wildness, the zoo exhibit should reflect as complete an ecosystem as possible. Exhibits should display not only animals but environments, complete with multiple species of flora and fauna and illustrations of ecological interactions. Not only is this a better representation of nature; it also reminds the visitor of the interdependence among animal and plant species. In unusual cases, such ecosystems could also incorporate human activity. For example, some environmentally conscious elephant centers in Thailand, recognizing the historically strong emotional connections between elephants and mahouts, house the elephant caretakers on site and incorporate them into the interactions that visitors can experience (figure 4.1).

FIGURE 4.1. The relationship between elephants and their human keepers is recognized as an important part of elephant conservation at some conservation centers in Thailand, and the mahouts help the visitors engage appropriately with the elephants.

Source: Photo by the author.

Another model for zoos increases the emphasis on the landscape, so that even if the animals are not visible the visitor feels that they have been able to experience the wild. Safari parks transport visitors through a "wild" landscape in which they can see animals living in what appears to be (but typically is not) their natural state. Although some people distinguish between zoos and safari parks, the line between them is permeable. In Ohio, for example, an area called the Wilds is associated with the Columbus Zoo and presents wild animals who inhabit broad (though fenced) landscapes rather than small enclosures. Visitors are transported through the landscape in shuttle buses rather than walking or driving their own car. The animals, which include giraffes and rhinos, are clearly not in their natural habitat, but the visitor can maintain the illusion that they are living "naturally." Interestingly, a survey of visitors to the Wilds showed that people who rated the landscape as more wild also gave higher ratings of enjoyment to their visit, said they felt a greater connection to the animal, and were more supportive of environmental conservation.[19] They were also more likely to talk to the people they were with about their visit, suggesting perhaps that a wild(ish) landscape can be effective in evoking social engagement. This is important because social interactions can solidify and strengthen the impact of an emotional experience at a conservation center.

In addition to experiences that highlight connection and emphasize ecosystems and landscapes, a third possibility for zoos would be to exploit new ways of conveying information, including technology. Many zoos already effectively utilize streaming video, and some are trying novel new ways to engage visitors with games or other apps that intersect with the live zoo environment. People want to be able to see the animals at the zoo and ideally to engage with them; this may help satisfy the goal of connection, but it can be at odds with the idea of presenting a wild landscape, in which the animal may be obscured by foliage. Technology is already used to help people encounter the animals that may not be directly visible, thus addressing the needs of both the animals and the zoo visitors. Given the increasing presence of technology in people's lives, zoos may be able to have the biggest impact by exploiting it rather than resisting it. Art can also be a way of magnifying the zoo experience; poetry or visual art can encourage visitors to take new perspectives on the animals and plants they have observed (as discussed by Alison Deming in the previous chapter).

Finally, zoos may be able to strengthen people's connections to nature by giving them more practical ways to engage with it. Zoos tend to provide information about conservation needs but are less likely to suggest specific behaviors that visitors can engage in. But many people say that their reason for not doing more for conservation is that they don't know what to do.[20] Skills training can be important for people who already support environmental conservation, and it can also provide an important boost in efficacy. If people recognize their own potential to help conservation efforts, they may feel more connected to the goal.

WHAT IS THE WILD?

The idea that technology can be part of a zoo visit may lead to the question: what is the future of wildness? Although many people's attitudes, as well as many governmental policies in the United States (e.g., the Wilderness Act), are based on a strict separation of wild and tame,[21] this approach is becoming harder to maintain because there are no more wild areas that have not been affected by human activity. However, there are advantages to moving past this duality, which artificially separates humans from the rest of the world. In fact, the value for wilderness is culturally grounded. Many cultures, particularly indigenous ones, do not have a word for the concept, recognizing instead that humans and other species live together as a community.[22] In our modern, densely built and inhabited world, this does not mean that we don't need to have landscapes without human residents. It may, however, suggest that we recognize that humans have a role in constructing and managing those landscapes—as well as the ways in which we have the potential to degrade them. When we have places that are perceived as wild—such as zoos or even parks or nature reserves—we acknowledge that humans have been part of their creation.

One interesting and valuable consequence of acknowledging the human influence over wild landscapes is that it allows greater room for the possibility of empathy in managing those landscapes. Some rewilding projects have been criticized for allowing and even inadvertently promoting the suffering of nonhuman animals.[23] From a "hands-off" position there is nothing to be done when large numbers of animals starve or are brutally preyed upon by others. If humans are allowed to intervene, they can try to maintain an ecosystem that experiences fewer dramatic fluctuations and thus avoids massive die-offs. Allowing some evidence of human influence as part of wild landscapes may also help us develop coherent policies regarding things such as cell phone use or accessibility to people with disabilities in wilderness areas (not that those decisions are easy).

Rejecting the duality between wild and tame, natural and human, may also mean rejecting the distinction between real and simulated. Although zoos are artificially constructed landscapes, they present a certain representation of nature to the visitor that is likely to become increasingly important as access to wild nature is reduced. This representation can both respond to and shape visitor expectations for wildness. I would argue that zoo displays should attempt to address several interconnected goals. First, as described earlier, they should present the animal as part of an ecosystem and acknowledge that people also exist in that ecosystem. That doesn't mean allowing people to live next to the animals, but it does mean being candid about the human interventions that are required to maintain the animals and environments in a healthy condition. The second goal is to emphasize that the animals have value—not just economic value but humanistic value; that people care about them, in other words. (The animals may also be allocated intrinsic value, unrelated to any human attitudes, but that is a complex argument for a zoo to make while keeping them captive.) Both the social norms surrounding what is valued and personal connections to the animal can promote visitors' tendency to personally assign value to the animals. A third goal is to encourage the visitor to recognize their own role in the system: how their behaviors can put species and ecosystems at risk and, on the other hand, how they can act in ways that promote ecosystem well-being.

As described earlier, connection includes both a cognitive appreciation of interdependence and an emotional response to the animals. In trying to encourage a sense of connection as well as providing experiences of the wild, zoos confront a challenging task of emphasizing both similarity and difference. A well-constructed zoo visit will emphasize that the animals are not simply objects of the visitor's gaze but fellow members of the natural community, while also providing a reminder that nature is not entirely in the human domain; that not all animals are domestic, and these animals inhabit very different lives from those of humans; and that there is still value in the idea of the wild. Such a perspective could satisfy visitors' desires to experience the wild while reminding them of their own role in preserving it.

NOTES

Author's note: Work on this chapter benefited from a fellowship at the Fondation Maison des Science de l'Homme in Paris and from a FIAS fellowship at the Paris Institute for Advanced Study (France). It has received funding from the European Union's Horizon 2020 research and innovation program under the Marie Skłodowska-Curie grant agreement No 945408 and from the French state program "Investissements d'avenir," managed by the Agence Nationale de la Recherche (ANR-11-LABX-0027-01 Labex RFIEA+). It was also helpfully guided by a workshop at the Arizona-Sonora Desert Museum and by feedback from the participants, especially the editors of this volume.

1. Patricia G. Patrick and Susan Caplow, "Identifying the Foci of Mission Statements of the Zoo and Aquarium Community," *Museum Management and Curatorship* 33 (2018): 120–35.
2. Eric Jensen, "Evaluating Children's Conservation Biology Learning at the Zoo," *Conservation Biology* 28 (2014): 1004–11. See also Kara K. Schroepfer et al., "Use of 'Entertainment' Chimpanzees in Commercials Distorts Public Perception Regarding Their Conservation Status," *PLOS One* 6, no. 10 (2011): e26048.
3. Susan Clayton, John Fraser, and Carol D. Saunders, "Zoo Experiences: Conversations, Connections, and Concern for Animals," *Zoo Biology* 28, no. 5 (2009): 377–97.
4. Coral M. Bruni, John Fraser, and P. Wesley Schultz, "The Value of Zoo Experiences for Connecting People with Nature," *Visitor Studies* 11, no. 2 (2008): 139–50; Matthias Winfied Kleespies et al., "Connecting High School Students with Nature—How Different Guided Tours in the Zoo Influence the Success of Extracurricular Educational Programs," *Frontiers in Psychology* 11 (2020): 1804.
5. Tiffani J. Howell, Emily M. McLeod, and Grahame J. Coleman, "When Zoo Visitors 'Connect' with a Zoo Animal, What Does That Mean?," *Zoo Biology* 38, no. 6 (2019): 461–70.
6. Susan Clayton, "Attending to Identity: Ideology, Group Membership, and Perceptions of Justice," in *Advances in Group Processes: Justice*, ed. Karen A. Hegtvedt and Jody Clay-Warner (Bingley: Emerald, 2008), 241–66.

7. Gladys Barragan-Jason et al., "Human-Nature Connectedness as a Pathway to Sustainability: A Global Meta-Analysis," *Conservation Letters* 15 (2022): e12852, https://doi.org/10.1111/conl.12852.
8. Clayton, Fraser, and Sanders, "Zoo Experiences."
9. Neil Carr, "Ideal Animals and Animal Traits for Zoos: General Public Perspectives," *Tourism Management* 57 (2016): 37–44.
10. Richard G. Walsh, and John B. Loomis, "The Non-Traditional Public Valuation (Option, Bequest, Existence) of Wilderness," in *Wilderness Benchmark 1988: Proceedings of the National Wilderness Colloquium*, comp. Helen R. Freilich (Asheville, NC: Southeastern Forest Experiment Station, 1989), 181–91; Adam C. Turner et al., "Comprehensive Valuation of the Ecosystem Services of the Arctic National Wildlife Refuge," *Natural Areas Journal* 41, no. 2 (2021): 125–37.
11. E. O. Wilson, *Biophilia* (Cambridge, MA: Harvard University Press, 1984).
12. Juliet Clutton-Brock, "The Wild and the Tame," in *Wildlife Conservation, Zoos, and Animal Protection: A Strategic Analysis*, ed. A. Rowan (Boston: Tufts University Center for Animals and Public Policy, 1995), https://www.ebookmakes.com/pdf/wildlife-conservation-zoos-and-animal-protection/.
13. Ben A. Minteer and Christopher Rojas, "The Transformative Ark," in *A Sustainable Philosophy: The Work of Bryan Norton*, ed. Sahotra Sarkar and Ben A. Minteer (Springer, 2018), 253–71; Bryan G. Norton, *Why Preserve Natural Variety?* (Princeton, NJ: Princeton University Press, 1987).
14. Sean P. Goldy and Paul K. Piff, "Toward a Social Ecology of Prosociality: Why, When, and Where Nature Enhances Social Connection," *Current Opinion in Psychology* 32 (2020): 27–31.
15. Susan Clayton, "The Psychology of Rewilding," in *Rewilding*, ed. Nathalie Pettorelli, Sarah M. Durant, and Johan T. du Toit (Cambridge: Cambridge University Press, 2019), 182–200.
16. Brendon M. H. Larson, Bob Fischer, and Susan Clayton, "Should We Connect Children to Nature in the Anthropocene?," *People and Nature* 4 (2022): 53–61.
17. Stephen Kaplan, "The Restorative Benefits of Nature: Toward an Integrative Framework," *Journal of Environmental Psychology* 15, no. 3 (1995): 169–82.
18. Susan Clayton and Khoa D. Le Nguyen, "People in the Zoo: A Social Context for Conservation," in *The Ark and Beyond: The Evolution of Zoo and*

Aquarium Conservation, ed. Ben A. Minteer, Jane Maienshein, and James P. Collins (Chicago: University of Chicago Press, 2018), 204–11.

19. Lila B. E. Gibson, "The Short- and Long-Term Impacts of a Safari Park Visit on Guests' Environmental Attitude and Behaviors," Bachelor's thesis, College of Wooster, 2020, https://openworks.wooster.edu/independentstudy/9149.
20. Susan Clayton et al., "Confronting the Wildlife Trade Through Public Education at Zoological Institutions in Chengdu, China," *Zoo Biology* 37, no. 2 (2018): 119–29.
21. Clayton, "The Psychology of Rewilding."
22. Enrique Salmón, "No Word," in *Wildness: Relations of People and Place*, ed. Gavin Van Horn and John Hausdoerffer (Chicago: University of Chicago Press, 2017), 24–32.
23. Helen N. Kopnina, Simon R. B. Leadbeater, and Paul Cryer, "Learning to Rewild: Examining the Failed Case of the Dutch 'New Wilderness' Oostvaardersplassen," *International Journal of Wilderness* 25, no. 3 (2019): 72–89.

5

"WILD" THROUGH AN AMERICAN INDIAN HISTORICAL ANALYSIS

KELSEY DAYLE JOHN AND REVA MARIAH SHIELDCHIEF

Like all things, our tribal understanding of the word *wild* is determined relationally. The simple "English" dictionary definition of wild differs significantly from the authors' relational understanding, crafted historically in a network of policies, resistances, interpersonal anecdotes, and ultimately tribal languages to which we have been exposed. The process of understanding a word requires remembering, reflection, and dialogue with our relations as we weigh our cultural and societal responsibilities. We both are American Indians but come from distinct Tribal Nations with distinct experiences and constructs of identity built within contemporary *American* society.

Regardless, relationality for the purpose of this chapter is understood as concerning both our homelands and, when applicable, our current land bases (i.e., Tribal Nations who were removed from the lands of their creation and placed in Oklahoma). Fundamental to those relations are our relationships to, and agency with, all *persons* that live upon those habitats. When considering practices meant to rewild and/or making zoos "wilder," we urge readers to consider the historical context through which the concept of "wild" emerged and how those historical contexts influence the theory and practice of wilding in the current moment. We also urge readers to remember there is a power dynamic inherent in one's ability to declare or claim wildness for other beings.

WILD

Kelsey (Navajo)

When I remembered my education around *wild*-ness, I remembered my family playfully pushing back on the term using humor and intellect. There is something tongue-in-cheek about my family's use of the word *wild*. Growing up, we were often in some relation to that which is labeled *wild*. But I never understood *wild* to be negative; it seemed to me that *wild* meant free. Mostly, I learned this from mustangs. We'd joke about our *wild* mustangs, even though these were the horses I grew up riding, and I still find it much easier to communicate with them than with domestic horses—and much easier to communicate with than human domesticates.

In my twenties, I heard another Indian use the word *wild* playfully and for some reason it stuck in my memory. At a conference, two friends were meeting after being apart for some time, both Native women. One screamed, "Hey there! My favorite *wild* Indian!" This little salute meant something to me, made me pause. A quippy greeting with a bouquet of historical meaning, humorously pushing back on the idea that we, Indian women, are often called *wild*. I thought, I like this woman. She gets it. We are *wild*, in the best kind of way.

I see "wild Indians" like other phrases or words that communities reclaim. Historically, the words used to categorize and deplete persons from their personhood tend to be reclaimed by those same communities. This is why we can jokingly call one another *wild*; the joke is the resistance. I made the connection later; my family always used this phrase to describe the ridiculousness of the idea that someone or something is truly *wild*. Like other tropes my family critiqued through humor, I have no memory of *wild* used in a derogatory way; it was always playful, representing freedom and our relationship to freedom—something the white settler world seemed to lack. In my first published piece, I mention that my dad "breaks *wild* mustangs." I do this to juxtapose *wild*/domestic, patriarchy/feminism, and me/my dad in the work we do with our community. Even though the piece had nothing to do with horses, I always had an idea that relations with these so-called wild beings made my family different.

Mariah (Skidi-Pawnee and Tohono O'odham)

When asked to consider the terms and conceptualizations of wild and wildness as it pertains to our Tribal Nations and our nonhuman-person relations, I excavated the terms as they apply to me and my Tribal (Skidi-Pawnee and Tohono O'odham) upbringings, particularly my relationships with, and relationality to, animal-persons. I had to ask myself *if* within my family and, by extension, my tribal culture's lessons and rearing the terms "wild," "wildness," or "wilderness" were used. If so, in what context were the terms used and to whom (land, animal, or human-persons) were the terms applied?

Thinking, as I often do, on my drive just west of the Tucson mountains to feed my horses, it struck me that the concept of wild animals and/or wild landscapes was not a part of the conversational lessons my father or *Uppitt* (Skidi-Pawnee for "my grandfather") ever had with me regarding places or animal-persons. Nor had they come from my *hu'li* (a female speaking Tohono O'odham for "my grandmother") and mother, who gave much less intentional teachings. Making that rather remarkable observation, I wondered if Kelsey had a similar experience growing up. Conversing, we began a dialogue about our tribal languages and if "wild" and "wilderness" existed within our personal tribal communities. To be sure of what I felt to be true about my Skidi understanding, I conferred with my father. The concept of "wild" did not describe animal-persons, plant-persons, or the contours of the land we lived upon or happened to be visiting.

Rather, *wild* was a word we used for "fearless" or "bold." Most times it was used, in *English*, for the admirably dangerous behaviors of human-persons. When speaking, *wild* was blunt and cutting honesty. Acting wild could be any act that required guts more than skill. *Wild* was also admirable. It was, as Kelsey expressed, an action or attitude that reflected a person's choice to be bold, free of more careful tribal/societal/community constructs of discourse or action.

When I spoke with my father about wild and wildness, we reminisced about what *Uppitt* would say about such things. We concluded that animal-persons, such as a bear, a powerful and adamant being especially upon his own landscape, were respected, and we Skidi-persons would heed what such a person would communicate. There is a relationship between me as *Skidi*-person[1] and *Kuruks* (Bear). The way that relationship is expressed is not the question. I was taught to be mindful and respectful of a bear-person's agency in any space, whether in the Rockies, Madera Canyon, a circus, or in a zoo. Simply put, while I understand the usage of "wild," "wildness," and "wilderness" within Eurowestern society, I have determined not to use those definitions here. Instead, I humbly choose to express what I know as a Skidi-Pawnee and a Tohono O'odham.

LANGUAGE

We have discussed and made a tacit agreement to speak of "animals" as Persons. We kicked around the idea of referring to those four-legged, two-legged, winged, or finned (as well as those leafed and vined, stalked, and trunked) as "beings" but felt that the agency these relatives hold needed to be recognized. In the literature, animal persons are referred to as nonhuman, more-than-human, and animal. However, we find that the mutual recognition of these diverse beings as Persons is most in line with their positionality in our respective worldviews. You will find that we use the word Persons because in our intricate web of relationships and relationality, we are simply declaring a truth: animals are people in the same way that humans are people.

HISTORICAL AND SOCIAL CONTEXT OF THE AMERICAN INDIAN

It's common in pieces written for mostly non-Native audiences to find the incessant idea that Nativeness exists in the persistent trope of *Noble Savagery*.[2] The noble savage is the idea that Native persons are noble, reverent, and romanticized. Yet, at the same time, they are also savage, despicable, and backward. The "noble" and the "savage" are married in the American myth and exist in a dichotomy woven so tightly together that it is nearly impossible for an image to be read as only noble or only savage; it's always some poisonous mix of the two. As we discuss and describe the word *wild*, we ask you the reader to think critically about the *noble savage* trope and challenge yourself to do the uncomfortable work of pushing back on this narrative in your own digestion of these words. We want to draw attention not to the overemphasized myth that Natives are somehow closer to nature and therefore more primitive but to a truth inherent in tribal worldviews about instructions and stories embedded in layers of relationality.

To put this into context, let us take a deeper dive into the philosophy that pervaded the major powers that began colonizing what is now the United States and their unifying Christian epistemology.[3] While it may feel like a strange and cumbersome fit, we are suggesting that the underlying and pervasive cultural and societal ontology of Eurowestern and therefore EuroAmerican settlers hinge upon Christian morals, values, and worldview. The *noble savage*, then, is not only an oxymoron of sorts, but it behooves us also to take into account that a predominantly Christian worldview expressed in the Bible includes within its own creation story and Old Testament accounts the dominion of "Man" over women, animals, and nonbelievers. The noted difference, and one applicable here as we discuss the concepts of *wild* and *wildness*, is the predisposition to view the earth and its inhabitants as "Man's" dominion. Is it then strange that Christian Eurowestern settlers were at a loss to aptly "name" and conceptualize tribal nations' members with whom they were creating political relationships? They sought to *civilize* via policies of removal, boarding schools, reservations, and allotment. All are attempts to Christianize.

Examples of EuroAmerican society's romantic and misleading characterizations of tribal people, such as the *noble savage*, can be easily witnessed in the popular literature of the time. Novels and stories such as Cooper's *The Last of the Mohicans* (1826) and Hawthorne's "Young Goodman Brown" (1835) are two earmark examples wherein the image, and trope, of "Indian" as *noble savage* played out. Of note, both of these texts predate the Civil War and the expansion of the American West through Manifest Destiny that followed. Everything we are suggesting here has already been written about and debated ad nauseam across many disciplines.[4] The overarching religious ontology of EuroAmerican settlers and policy makers had a part to play in the creation of removal, reservations, and allotment, all of which were intended to corral and restrict the space in which a tribe might live, hunt, and gather. U.S. policy grounded in European ontologies expressed a Christian understanding of Dominion not just over Indian humans (utilizing the idea of the *noble savage*) but over the American West, plains, mountains, rivers, lakes, bison, horse, wolf, etc.

In contrast, the *wild*, heathen, or non-Christian worldview of peoples of Tribal Nations is, as we've described, one that seeks to clarify Tribal and Native perspectives. Vine Deloria Jr. speaks of these relationships in *God Is Red*, writing, "In the religious world of most tribes, birds, animals, and plants compose the 'other peoples' of creation."[5] Just before this quotation, he provides an example of the deeply religious/spiritual space in which both human-persons and animal-persons live as well as the agency of animal-persons:

> It may indeed be the starting point of the Great Race that determined the primacy between two-legged and four-legged creatures at the beginning of the world. Several mountains in New Mexico and Arizona mark places where the Pueblo, Hopi, and Navajo peoples completed their migrations, were told to settle, or where they first established their spiritual relationships with bear, deer, eagle, and *other peoples* who participate in the ceremonials.[6]

What we can ascertain is that the relationships between ourselves as human-persons and what is *wild* are all formulated as a part of the origin of a Tribal Nation's identity. The identity of our Tribal Nations is woven into our creation stories, our songs, ceremonies, prayers, and other sacred information not accessible in decontextualized spaces such as academic publications. Sharing, in the Native, or tribal, context, *is a relational process*. The relational process involves breaks and stops and refusals to share everything in a linear way,[7] making the writing process a challenge to construct linearly.

The ironic dichotomy of the noble savage resonates within federal Indian policy and its relationship to both research and cultural representations of the "Native," (i.e., members of Tribal Nations). Some forty years after the Civil War was won and lost, Manifest Destiny had spread across the plains, wherein Tribal Nations were pacified and corralled onto reservations. There was a push by American anthropologists to rush to the West to record and photograph human-persons of Tribal Nations before we lost our *Indianness*. Many had already been pushed off our original homelands and relocated hundreds if not thousands of miles away from what we knew as home. Animal and Plant-persons understood as close relatives were left behind, forcing us to establish new relationships in new landscapes.

Interestingly, there was concern that the Western European culture of the United States would soon "scrub the savage from the Indian." Ironic as that seemed, that was exactly what army captains, U.S. legislators, Supreme Court rulings, boarding schools, and removal aimed to do—that is, to civilize the *wild* Indian. However, to keep the Indian from vanishing, there are social and political investments in a whole new debate regarding the relationship between the U.S. federal government and tribal nations, which began as a response to the failed policies of the General Allotment Act (Dawes Act) and boarding schools such as Carlisle and Chilocco. The Glass Door policy illustrates this troubling dichotomy, asking the ongoing question of which of two options were better for Indian Tribes: to be separated and cocooned in/on their own reservations to develop and evolve at their own pace without influence from Eurowestern communities/cities[8] *or* for the U.S. federal government to adequately supply jobs and training as well as goods and services to tribal-persons.

Our goal in highlighting this history is to draw a connection between Indian policy and the development of institutions meant to display/preserve Indian-ness (culture, people, or objects), both of which emerge within the same twenty-year period. We argue that the idea of "vanishing Indians" was the driving force behind fairs, museums, and displays that were meant to preserve those vanishing Indians. This same fear contextualizes other institutions and policies happening in Indian country during the same time period. Consequently, the "vanishing Indian myth" must be understood as a derivative of the *noble savage* myth—where preserving the nobleness always comes with the belief that the noble Indian remains always savage. Conserving and preserving is met with destruction, violence, and othering. Examples of this practice include Wild West shows (e.g., Buffalo Bill), museums, and zoos, all of which became extremely popular in the same generation. We read the actions of preservation to be deeply connected to ideas of destruction and the advancement of progress.

ZOOS

Historically, zoos were spaces and institutions constructed to represent and reaffirm a believed-to-be-"natural" division between *wild*/domestic animal-persons. Gary Paul Nabhan's chapter in this volume also critiques the foundation of zoos as spaces that enclosed and isolated beings who were once in relation; he even suggests that zoos have contributed to extinction through isolation. At the same time, zoos have engineered a separation observed in human communities between "Indigenous/Native" and other (historically human and more contemporarily animal) domestic beings and spaces that exist in a bilateral relation to *wild*/nondomestic beings.[9] Walter Putnam explains that zoos positioned the "close to nature" Native body outside its *wild* context and put those bodies on display for the domestic public. He writes: "Colonial subjects exhibited in zoos or fairs were framed as objects taken from an imagined environment and put on display for their very nature. According to the racial paradigm of the day, natives were supposed to embody nature; they were supposed to be natural, even in the most unnatural of settings."[10]

The positioning of a "close-to-nature" Native in a designated non-*wild* space gave direct access to imagery that reaffirmed the notion of Indians as *wild*, more natural, and more primitive. Constructed to reaffirm the relationship between *wild*/domestic beings is the erection of separate spaces where one person is on display and another person gazes. Also related to the narratives about *wilderness*/nature is the persistent idea that both Native peoples and Native spaces are disappearing or will no longer be *wild* but eventually "domestic." We can see this in the discourse on conservation, where spaces are conserved insofar as the public believes that nature and *wilderness* are disappearing.[11] Ironically, the creation of preserved natural spaces like national parks required the forced removal of Indigenous persons (human and nonhuman).[12]

Other public institutions, including museums, reflect the colonial worldview through which they were created. These spaces are important because they shape the public and historic narrative of people and relations but are most times constructed through a worldview that reflects a Christian ideal of domination and the idea that Natives are *noble savages* on display. In a piece on decolonizing museums, Elisa Schoenberger examined a number of museums undergoing decolonial processes in conversation with Indigenous experts in the field. She summarizes the idea of decolonization as it pertains to museums, writing:

> It's not just about inviting indigenous and other marginalized people into the museum to help the institution improve its exhibitions; it's an overhauling [of] the entire system. Otherwise, museums are merely replicating systems of colonialism, exploiting people of color for their emotional and intellectual labor

> within their institutions without a corollary in respect and power.[13]

The process of decolonization, experts argue, is a process of dismantling the narratives that undergird systems built upon the dominant notions of relationality, labor, and history. We argue that *wilderness* and *wild* also need to be contextualized within our tribal spaces and worldviews.

Aside from the destructive context that accompanies the meaning of *wild*, we believe that these contextual characterizations influence vast sets of relations between human and nonhuman persons. To illustrate this pervasiveness, one could choose from a collection of human interventions specifically in the Southwest region that have adversely affected the nonhuman world. There are countless projects developed within the nexus of settler colonialism where the idea of a *wilderness* and *wild* Indian actors demonstrate the ongoing supremacy of civilization. However, it is not our intention to provide a comprehensive argument about colonial land projects in the Southwest but to make the connection between human-engineered landscapes (occupied by Indigenous human and animal nations) and the logic of settler colonialism often represented and retold in the nature/culture dichotomy of zoos.

In her chapter in this book, Amanda Stronza argues that there is a human-orchestrated component to wild spaces and that wild environments are engineered by humans for entertainment, confinement, and hierarchy. That is the idea of a tameless, uncivilized nature that must be converted into a cultural, civilized, American standard. What persists in any zoo, museum, and research project is the idea that some are *wild* and others are not. As we provide a critical lens to the social construction of *wild* and *wilderness*, we hope to create new narratives while critiquing and dismantling mythic narratives about *wild* beings and *wild* spaces.

THE IRONY OF "WILD"

When we think of the word *wild* as the rallying concept through which we might seek to "re*wild*" zoos and animal habitats, we first thought about *wild* in our communities meaning "free." But in order to understand why this word is reimagined in our Native communities, we must first understand how *wild* has historically marked some of us as unfree to pursue relations while others are *privileged* to be marked as "tame" or un*wild*. The nomenclature is challenging because today *wild* colloquially means "free" yet has been the primary marker for making Indigenous persons (human and animals) unfree to pursue our traditional relations in our communities. This unfreedom comes from being marked as the negative version or savage side of *wild*, while settlers enjoy the noble side of *wild*, a *wild* that is mysterious, romanticized, and free. Natives have a long history with the word *wild*; it has been used against us, but eventually we reclaim it in our communities, much like the stories we shared at the beginning of this chapter. Settler societies enjoy the privilege of never having to encounter a savage side of *wild* and remain free to define and understand *wild* as mystical, magical, and happy. They never had *wild* used as weapon against them, nor were they equated with the *wild* to dispossess them of personhood.

RELATIONSHIP, AGENCY, AND "WILD-NESS"

Following our discussion connecting the historical context for American Indian communities to the more current work in decolonizing museums and zoos, we wish to close this chapter by raising several questions about moving forward within the work of human/animal person relations. For us, making a "*wilder* kingdom" must recognize and invest in relationships with the nonhuman world, evaluating those relationships for their participation in structures of settler colonialism.[14] To address this, we believe there must be a central effort to dialogue, collaborate, and craft mutually beneficial relationships with tribal nations—human and nonhuman alike. Just as Deloria reminds us, animals are communities comprising their own nations.

Ben and Harry put a series of provocative questions to the authors at the outset of this project. We'd like to close this chapter with our own open questions about the wild and zoos as we look into the future:

- How can we offer an understanding of *wild* less rooted in the history of settler colonialism?
- How might we think together with more-than-human persons about relations inherent in spaces labeled as *wild* and un*wild*?
- How do our practices center the personhood and being of animal persons?
- Can zoos perpetuate good relationships with animal persons? How do we define "good relations" between human/animal?
- What do humans desire from *wilder* kingdoms? How can we move beyond the desire to make some *wild* and others domestic?

NOTES

1. There is irony to be found in the fact that, in my band's Cadoan-based language "*skidi*" is the word for "wolf." In fact, Skidis were known for scouting while in the guise of wolf skins, acting and appearing as wolves, to watch enemy encampments.

2. For a complete discussion of the noble savage see Phillip Deloria, *Playing Indian* (New Haven, CT: Yale University Press, 1998); and Vine Deloria Jr., *God Is Red: A Native View of Religion*, 30th anniversary ed. (Golden, CO: Fulcrum, 2003).
3. Though there were many European nations who came to the North American continent, we are referring to the three European powers who have had continuing influence: Spain, Britain, and France.
4. Ad nauseam: George Pierre Castile and Robert L. Bee, ed., *State and Reservation: New Perspectives on Federal Indian Policy* (Tucson: University of Arizona Press, 1992); Vine Deloria Jr. and Clifford Lytle, *The Nations Within: The Past and Future of American Indian Sovereignty* (Austin: University of Texas Press, 1998); Vine Deloria Jr. and David E. Wilkins, *Tribes, Treaties, and Constitutional Tribulations* (Austin: University of Texas Press, 1999); and Janet A. McDonnell, *The Dispossession of the American Indian* (Bloomington: Indiana University Press, 1991).
5. Deloria, *God Is Red*, 274.
6. Deloria, *God Is Red*, 273
7. For more on refusals see Audra Simpson, "On Ethnographic Refusal: Indigeneity, 'Voice,' and Colonial Citizenship," *Junctures: The Journal for Thematic Dialogue* 9 (2007): 67–80.
8. Lewis Meriam, *The Problem of Indian Administration. Report of a Survey Made at the Request of Honorable Hubert Work, Secretary of the Interior, and Submitted to Him, February 21, 1928* (Baltimore, MD: Johns Hopkins University Press, 1928).
9. Sara Shahriari, "Human Zoo: For Centuries Indigenous Peoples Were Displayed as Novelties," *Indian Country Today*, August 30, 2011, https://indiancountrytoday.com/archive/human-zoo-for-centuries-indigenous-peoples-were-displayed-as-novelties.
10. Walter Putnam, "'Please Don't Feed the Natives': Human Zoos, Colonial Desire, and Bodies on Display," in *The Environment in French and Francophone Literature and Film*, ed. Jeff Persels (New York: Brill, 2012), 57.
11. Sarah Jaquette Ray, *The Ecological Other: Environmental Exclusion in American Culture* (Tucson: University of Arizona Press, 2013).
12. David Truer, "Return the National Parks to the Tribes: The Jewels of America's Landscapes Should Belong to America's Original Peoples,"

Atlantic, April 12, 2021, https://www.theatlantic.com/magazine/archive/2021/05/return-the-national-parks-to-the-tribes/618395/.

13. Elisa Schoenberger, "What Does It Mean to Decolonize a Museum?," *Museum Next*, May 11, 2021, https://www.museumnext.com/article/what-does-it-mean-to-decolonize-a-museum/.
14. For more on animals and settler-colonial structures, see Billy Ray Belcourt, "Animal Bodies, Colonial Subjects: (Re)locating Animality in Decolonial Thought," *Societies* 5, no. 1 (2015): 1–11.

6

TOWARD A WILDER KIN-DOM

Why Zoos Must Focus More on Ecological Interactions (with Our Children and Other Biota) Than on Isolated Species

GARY PAUL NABHAN

As the world listened intently to the rhetoric at the COP26 Climate Accord congress in Glasgow in November 2021, a group of scholars and practitioners gathered at the Arizona-Sonora Desert Museum (ASDM) to consider the future of wildness, particularly in relationship to zoos. None of us gathered at that world-famous "museum of relationships" have any trouble with the notion that our world as we know it—Planet Desert—is in crisis. Yet few COP26 delegates likely realized that the zoos, aquaria, and botanical gardens touted as safety nets for the many species now endangered on the planet have fallen into financial and health crises since 2020. These institutional crises, particularly in relation to their conservation work, have been reported by the BBC, *National Geographic UK*, the *New Yorker*, and the *Washington Post*.[1]

But as early as 1990, the nature writer Barry Lopez suggested that zoos—including the one where we gathered three decades later—have been in an ethical crisis for much longer because of their regrettable *enclosure of the wild* in the world. Lopez said as much in a poem he wrote called "Desert Reservation":

I'd heard so much good about this place,
how the animals were cared for
in special exhibits. But
when I arrived I saw even
prairie dogs had gone crazy in
the viewing pits; Javelina had no mud to
squat in, to cool down; Otter was
exposed on every side, even in his den.
Wolf paced like a mustang,
tongue lolling and crazy-eyed,
unable to see anyone who looked like
he did—only Deer, dozing opposite in
a chain-link pen.

Near the end of his poetic critique—in fact, in the first and last poem the National Book Award–winning Lopez ever published, he wrote:

At night when management is gone,
only the night watch left,
the animals begin keening: now
voices of Wood Duck and
Turtle, of Kit Fox and everyone else,
Bear too, lift up like the bellowing
of stars and kick the walls.[2]

If these public institutions are to climb out of these short-term and longer-term crises, I feel we need to take a journey back in time to better understand their roots and the trappings inherent in their origins. As Paul Shepard suggested, we cannot overcome our current societal limits of embracing (or respecting) and conserving the wild unless we go back to understand (and undo!) the trappings that zoos inherited from the founders of these public institutions.[3] Then and only then do zoos, aquaria, arboreta, and botanical gardens have a chance to venture out on a new bold trajectory rather than doing business as usual. And so, I'd like to take you all on a little journey, which may remind us that

- we need to remember why zoos, aquaria, and botanical gardens originally developed to enclose, display, tame, and simplify wild nature so that we can use that memory to chart a course toward a more resilient and needed role in the future;
- we need to acknowledge their historic shortcomings in isolating plants and animals from one another (and from us!) and how ASDM played a key historic role among all zoos by reintegrating them in outdoor exhibits of ecological interactions; and
- we need to reaffirm their key role in offering children direct (and at times "risky") contact with truly wild plants and animals, instead of merely offering safe hyperrealities.

Should we overcome our collective amnesia about these three points, we may find fresh ways to get more people to venture into the raw, improvisational theater where an evolutionary play is proceeding on a living stage called the Sonoran Desert, which is rapidly becoming the laboratory for the future on Planet Desert. We may then have a chance to engage them in the wild process that I call the trans-situ conservation of *holobionts*: truly wild plants, animals, and microbes interacting in ways that allow them to further adapt to changing environments to ensure their survival. Without such ecological interactions, it can be argued that captive-bred animals as well as horticulturally tamed plants have lost what it means to be wild. There is also the risk that their descendants will not survive in the wild even if reintroduced to complex "natural" ecosystems where enclosure does not inhibit such interactions.

So let us begin our journey near the beginning of the institutionalization of fauna and flora, early in the Industrial Revolution. One might start with Peter the Great's Museum of Anthropology and Ethnography (the Kunstkamera), which celebrated its three-hundredth anniversary in 2014. By 1716, Peter had begun to showcase his notion of the diversity of life on Earth by displaying a collection of more than two thousand pickled specimens made for the study of human anatomy and morphology, as well as a few "living specimens" of "human races," as they were known at the time.

Peter the Great seemed obsessed with deformed, stillborn fetuses, which medical biologists might now consider to be lethal genetic mutations. Peter also purchased more than 1,200 specimens of small mammals, birds, reptiles, amphibians, and marine and terrestrial invertebrates, then went on several multicountry expeditions to collect more on his own. Again, his own "trophies" and "purchases" of wild, other-than-human lifeforms disproportionately represented animals with deformities or monstrous organs, in his vain attempt to "capture" the gamut of wild nature on this planet for public display.[4] Although we might now consider it a "museum of unnatural history"—for it failed to provide any ecological, evolutionary, or genetic context to understand all these deformations—Peter's project is how the Occidental tradition of natural history museums was jumpstarted.

As the Industrial Revolution ramped up the destruction of natural habitats and their attendant flora and fauna in Europe, other institutions emerged to make up for those losses, often without addressing any of their root causes. The first widely read book of natural history—by Gilbert White—was initially published in 1789 but became the rage in the nineteenth century, going through three hundred editions. It became the first (and in some cases, the last) account of many species once common in the British countryside now though to be extinct.[5] Not long after White brought recreational nature watching into fashion, Kew Gardens opened its glass houses in metropolitan London to the public in 1840,[6] and the first "scientific zoological garden"—the ZSL London Zoo—opened its caged menagerie to public viewing in 1848.[7] There is little record of public opinion in Great Britain that revealed any immediate angst over the *enclosure of the wild* during this period.

In essence, all three nature-derived institutions emerged not long after the onset of the Industrial Revolution in Great Britain (1740–1860), when an agrarian and handicraft economy was replaced by one dominated by machine manufacturing, highly polluting industries, and the loss of once-common game and plant life in their natural habitats. As access diminished to wild habitats and their untrammeled fauna and unpotted flora in the British Isles, demand for such experiences increased.

Nostalgia generates surrogates and helps us feel complacent with them and what they stand for. Cynically, we might say that the investment in such surrogates for nature followed the supply-demand economic model originally developed by Antoine Augustin Cournot, who first explained the theory in a book published in 1838.[8] Within thirty years, the supply-demand trope was broadly publicized by Alfred Marshall and became widely accepted by the formally educated elite.[9] In essence, as contact with unexpurgated nature became a scarce commodity in the British Isles, demand for recreational experiences of its remnants (if not for their conservation) increased, creating a lucrative market for nature surrogates.

In the United States, the same trend of concern emerged with its Industrial Revolution, on the heels of the "scorched earth" practices during the Civil War. Concerns about loss of wildlife and human contact with them were fostered by George Perkins Marsh's *Man and Nature* in 1864. The first zoological park opened in Philadelphia in 1874; the U.S. Botanic Garden and its Victorian Conservancy had opened earlier, in 1850.[10] As natural habitats were ravaged and the interactions among the flora and fauna were unraveled, Americans began to reify outstanding wildlife—particularly charismatic megafauna—*in isolation*, detached from their natural range of habitats and their many ecological interactions with the plants that offered them nourishment, medication, and shelter, as well the microbes and microinvertebrates in their guts and fur.[11]

Fortunately, a more holistic strategy of nature study gained headwind in the 1890s, but then floundered with the "Nature Fakers" controversy about the risks of anthropomorphizing animals in 1903.[12] Nevertheless, romanticized surrogates for wild nature and a focus on heroic individual animals dominated books and films for the "Bambi Generation" from the release of Walt Disney's movie *Bambi* in 1942,[13] on through the rest of the twentieth century, reiterating a reductionistic worldview. Since 1942, our children have been bombarded with the "Bambi mentality" in ads, cartoons, play apparel, stuffed animals, and toys, to the extent that cutesy anthropomorphized animals are deeply ingrained in the modernist fabric of our society.

Let us just say that we are well into the third century of parents in the Western industrialized world expressing concern that their children are "losing" touch with wild nature, a phenomenon that my friend Robert Michael Pyle called "the extinction of experience."[14] Today, children are more actively engaged with "nature imagery" in digital media even when they lack firsthand contact with wild nature.[15] While that may offer solace and marketing opportunities for wildlife conservation educators, the hackneyed representations of animals and plants either as horrifyingly dangerous or romantically cute abstractions lead many youths into a fascination with—if not an addiction to—the hyperreal.[16]

We are all aware that hands-on "experiential knowledge of nature" based on sensory observations of nature has declined among American youth over the decades,[17] just as frequency of visits to zoos and gardens has also declined in recent years. But are we aware of how serious this shift has been? And does that shift inevitably mean that children in industrialized societies have a limited if not a false sense of what "wild" may mean to others, one that may place them at real risk of misjudgment should they ever have a close encounter with the other-than-tame kind?

Let's look at three studies, summarized by Andrew Balmford and his colleagues in the pages of the prestigious journal *Science*.[18] One of them is their own study of "species knowledge" of 109 schoolchildren in the United Kingdom who were given flash cards of both "natural" and "synthetic" (i.e., video game) species. The other two are my own published collaborations cited by Balmford et al., one with Sara St. Antione and one with Stephen Trimble, which was updated in my more recent book, *Ethnobiology for the Future*.[19] These three studies all tend to confirm E. O. Wilson's "Biophilia Hypothesis" that humans have an innate desire to catalogue, understand, and engage with other lifeforms.[20] Yet, they also warn that as industrialization and urbanization reduce children's direct interactions with wild nature, their interest in engaging with a variety of living organisms is becoming redirected—or misdirected—toward human artifacts and media images.

> This possibility may have grave consequences for biodiversity conservation. Our findings carry two messages for conservationists. First, young children clearly have tremendous capacity for learning about creatures. Second, it appears that conservationists are doing less well than the creators of Pokémon at inspiring interest in their subjects: During their primary school years, children learn far more about [iconic characters in games like] Pokémon than about their native wildlife . . . being able to name less than 50% of common wildlife types [where they live].

Balmford and colleagues conclude that "evidence from elsewhere links loss of knowledge about the natural world to growing isolation from it. People care about what they know. With the world's urban population rising by 160,000 people daily, conservationists need to reestablish children's links with nature if they are to win over the hearts and minds of the next generation."[21]

RESTORING RELATIONSHIPS BETWEEN AND AMONG SPECIES (INCLUDING OUR OWN) WILL BE KEY

Before I offer any constructive critique that might help zoos, botanical gardens, and outdoor museums move forward to better educate children and better conserve biodiversity in a changing world, I first want to celebrate the pivotal role the Desert Museum has played in two realms:

1. Being among the first outdoor museums to reintegrate plants and animals in habitat-based exhibits that focused holistically on place-based biotic communities rather than on isolated taxa or on taxonomic groupings of animals—the reductionistic "lions & tigers & bears syndrome"
2. Being among the first to aid in species recovery and reintroductions of captive-bred animals back into current, former, or potential ranges by offering knowledge gleaned from diet, behavior, and health hazards and stressors to USFWS species recovery teams, a precursor of both *trans-situ conservation* and *assisted migration.*

Bill Carr—the visionary who cofounded the Desert Museum—was born at the height of the Nature Study Movement in 1902 and installed diorama exhibits for the American Museum of Natural History as a young man. He then pioneered outdoor natural history exhibits at Bear Mountain Trailside Museums in Highland Falls, New York, before falling ill and moving to Tucson. Both at ASDM and at the Ghost Ranch in New Mexico he developed his vision for outdoor naturalistic habitats for wildlife, an exhibit style that dominates zoos to this day. But have they become "safe surrogates" for urbanites who visit "hyperreal" exhibits in museums instead of engaging in "risky" wild nature on the edge of the city? That's what Timothy Luke concluded in 1997, contending that the Desert Museum was merely "imagineering Southwestern environments as hyperrealities"[22] that now work against true nature conservation. In Luke's opinion, "this institution began with a preservationist ethic and does still have some potential as a corrective device, but it has occasionally slipped toward a more negative presence, working ideologically to anchor real estate development, suburban consumerism, and industrial growth."

Today, thanks to zoo-based conservationists like the late George Rabb of Brookfield Zoo, many zoos and botanical gardens are humbly conceding that they are not necessarily where plants, animals, and microbial biodiversity are best "conserved"! At the onset of the Industrial Revolution, zoos and aquaria did harbor iconic individuals of a handful of species as habitats were lost, game was overhunted, and skies, soils and rivers were contaminated. *But was that true conservation of the wild?* Most of those surrogates for wild species died from enclosure, for they had been left with virtually no ecological interactions to sustain them.

On September 1, 1914, the last known passenger pigeon, a female named Martha, died at the Cincinnati Zoo (she was roughly twenty-nine years old, with a palsy that made her tremble). In about 1999, the last known wild Pele lobelia, *'ōhā wai*, died in its habitat after being propagated in vitro at the Hawaii Rare Plant Facility, which followed its resurrection from seed in the soil at the last site that this species was found in the wild.

Sadly, it is time to ask an unsettling question: Is it possible that zoos and botanical gardens have *sped up* the rates of extinctions of species and disruption of ecological interactions by keeping threatened species captive in isolation from their symbionts?

There is mounting evidence to suggest that these extinctions of "ecological interactions" through enclosure and isolation are the rule, not the exception.[23] Sun et al. report that Tibetan macaques translocated from the wild into captivity had reduced fungal diversity and increased abundance of gut fungal pathogens compared to wild individuals.[24] Allan et al. report that endangered Armagosa voles' gut microbes shift away from those dominant in natural habitats to entirely different ones in captive-bred animals.[25] Speer et al. report that microbiomes are integral to conservation of parasitic arthropods.[26] Finally, Liz Ecker and I have found that "captive" Thornber's fishhook cactus did not survive reintroductions without their original mycorrhizae because of an increased susceptibility to pathogens.[27]

Nevertheless, as Luke noted optimistically, zoos and gardens may not be inexorably destined to unintentionally produce conservation failures.[28] But to do otherwise, they would need to invest in microbial and invertebrate ecology to play effective roles in the conservation of ecological relationships. To do so, these institutions may need to staff more "lab girls" than "macho wildlife dudes" if they want to play a more effective role in true conservation of holobionts—that is, species *with* their associated biodiversity![29]

Let us remember that ex situ (beyond natural habitat) repositories like zoos, aquaria, arboreta, gardens, and seed banks still receive as much as 97 percent of the "conservation" (sic?) funds that might otherwise be available for in situ wildlife and plant conservation. A 2003 feature in *National Geographic* drove this point home: "David Hancocks, a former ASDM director with 30 years' experience, estimated that less than 3 percent of the budgets of these 212 accredited zoos go toward conservation efforts. At the same time, he pointed to the billions of dollars spent every year on hi-tech exhibits and marketing efforts to lure visitors. Many zoos not affiliated with the AZA spend nothing on conservation."[30] Two decades later, the situation is not much changed. As a consequence, we may be losing habitats that harbor multiple species at risk, while seducing the public into thinking that species can be "saved" one at a time, in isolation, within off-site repositories. But if we are to turn away from this paved-over route of control, comfort, and isolation, our zoos and gardens must become "museums without walls"—with financial "dietary restrictions" on how much they will consume *just* for self-perpetuation and promotion rather than for field and lab science in service to conservation in the wild.

In summary:

- Zoo and garden staffs must spend less time "on grounds" and veer onto the unpaved path *in the wild* to foster unmediated relationships with other species, accepting risks and uncertainties.
- They must be *more* than tourist destinations by helping their visitors become participants in ecological restoration in the broadest sense of the term.
- These institutions should redress the ratios within their budget of ex situ / in-zoo animal care versus in situ habitat protection, restoration, and collaborative conservation.

Most important, zoos and gardens must tell a new set of stories about themselves, the biota within and beyond them, and the people engaged in such work. We do not need any more avatars of the Great White Hunter, Marlin Perkins clichés derived from *Mutual of Omaha's Wild Kingdom* (1963–2018), or even echoes of Hal Gras from the early days of *The Desert Speaks*. Zoos and gardens must open their doors to a greater cross-section of cultures, genders, and ages in whom they employ, whom they place on their boards, how they gain their support, and for whom *in this blessed creation* they serve.

NOTES

1. Helen Briggs, "COVID-19: Funding Crisis Threatens Zoos' Vital Conservation Work," *BBC News*, October 1, 2020, https://www.bbc.com/news/science-environment-53938561; Manish Pandey, "Zoos Struggling in Pandemic as Group Says Government Hasn't Done Enough," *BBC News*, February 9, 2021, https://www.bbc.com/news/newsbeat-55996588; Karin Bruillard, "Shuttered Zoos Are Hemorrhaging Money, and They Want Federal Help for Endangered Species Work," *Washington Post*, June 8, 2020, https://www.washingtonpost.com/science/2020/06/08/coronavirus-zoos-aquariums/; Robin Wright, "Some Zoos, and Some of Their Animals, May Not Survive the Pandemic," *New Yorker*, May 18, 2020, https://www.newyorker.com/news/our-columnists/some-zoos-and-some-of-their-animals-may-not-survive-the-pandemic; Simon Ingram, "Zoo Crisis Deepens Amidst Second National Lockdown and 'Restrictive' Bailout Conditions," *National Geographic*, November 14, 2020, https://www.nationalgeographic.co.uk/animals/2020/11/zoo-crisis-deepens-amidst-second-national-lockdown-and-restrictive-bailout.
2. Barry Lopez, *Desert Reservation* (Port Townsend, WA: Copper Canyon, 1990).
3. Paul Shepard, "Post-Historic Primitivism," n.d., https://archive.org/stream/Post-historicPrimitivism/Post-historicPrimitivism_djvu.txt.
4. Maxim Germer, "Kunstkamera—Peter the Great's Museum of Malformation and Anthropology," https://lidenz.ru/kunstkamera-in-saint-petersburg/.
5. Jill Atkins and Warren Maroun, "The Naturalist's Journals of Gilbert White: Exploring the Roots of Accounting for Biodiversity and Extinction Accounting," *Accounting, Auditing, and Accountability Journal* 33 (2020).

6. Christophe Bonneuil, "The Manufacture of Species: Kew Gardens, the Empire and the Standardisation of Taxonomic Practices in Late 19th Century Botany," in *Instruments, Travel, and Science: Itineraries of Precision from the Seventeenth to the Twentieth Century*, ed. M.-N. Bourguet, C. Licoppe, and O. Sibum (London: Routledge, 2002), 189–215; emphasis mine.
7. Takashi Ito, *London Zoo and the Victorians, 1828-1859* (London: Boydell & Brewer, 2014).
8. Martin Shubik, "Antoine Augustin Cournot," in *Game Theory*, ed. J. Eatwell, M. Milgate, and P. Newman (London: Palgrave Macmillan, 1989).
9. Arthur Marshall, *Principles of Economics* (New York: Liberty Fund, 1890).
10. John Sedgewick, *The Peaceable Kingdom: A Year in the Life of America's Oldest Zoo* (New York: William Morrow, 1988); Elizabeth Anne Hanson, *Animal Attractions: Nature on Display in American Zoos* (Princeton, NJ: Princeton University Press, 2002); Karen Solit, *History of the United States Botanic Garden* (Washington, DC: Government Printing Office, 1993).
11. Alfonso Valiente-Banuet et al., "Beyond Species Loss: The Extinction of Ecological Interactions in a Changing World," *Functional Ecology* 29 (2015): 299–307.
12. Kevin C. Armitage, *The Nature Study Movement: The Forgotten Popularizer of America's Conservation Ethic* (Lawrence: University Press of Kansas, 2009); Ralph H. Lutts, *The Nature Fakers: Wildlife, Science, and Sentiment* (Golden, CO: Fulcrum, 1990).
13. Felix Salten, *The Original Bambi: The Story of a Life in the Forest* (Princeton, NJ: Princeton University Press, 2022).
14. Robert Michael Pyle, *The Thunder Tree: Lessons from an Urban Wildland* (New York: Houghton Mifflin, 1993); Mashashi Soga and Kevin J. Gaston, "The Extinction of Experience: Loss of Human-Nature Interactions," *Frontiers in Ecology and Environment* 14 (2016): 94–101.
15. Richard Louv, *Last Child in the Woods: Saving Our Children from Nature-Deficit Disorder* (Chapel Hill, NC: Algonquin, 2005).
16. Timothy W. Luke, "The Arizona-Sonora Desert Museum: Imagineering Southwestern Environments as Hyperreality," *Organization and Environment* 10 (1997): 148–63.
17. Louv, *Last Child in the Woods.*
18. Andrew Balmford et al., "Why Conservationists Should Heed Pokémon," *Science* 295 (2002): 2367.

19. Gary Paul Nabhan and Sara St. Antione, "The Loss of Floral and Faunal Story: The Extinction of Experience," in *The Biophilia Hypothesis*, ed. S. R. Kellert and E. O. Wilson (Washington, DC: Island, 1993), 229–50; Gary Paul Nabhan and Stephen Trimble, *The Geography of Childhood: Why Children Need Wild Places* (Boston: Beacon, 1994); Gary Paul Nabhan, ed., *Ethnobiology for the Future: Linking Cultural and Ecological Diversity* (Tucson: University of Arizona Press, 2016).
20. Edward O. Wilson, "Biophilia and the Conservation Ethic," in *The Biophilia Hypothesis*, ed. Kellert and Wilson, 31–41.
21. Balmford et al., "Why Conservationists Should Heed Pokémon."
22. Luke, "The Arizona-Sonora Desert Museum."
23. Jane Memmott et al., "The Conservation of Ecological Interactions," in *Insect Conservation Biology* (Oxfordshire: Proceedings of the Royal Entomological Society, 2007), 226–44.
24. Binghua Sun et al., "Marked Variation Between Winter and Spring Gut Microbiota in Free-Ranging Tibetan Macaques (*Macaca thibetana*)," *Scientific Reports* 6 (2016): 26035.
25. Nora Allan et al., "Conservation Implications of Shifting Gut Microbiomes in Captive-Reared Endangered Voles Intended for Reintroduction Into the Wild," *Microorganisms* 6 (2018): 94.
26. Kelly Speer, et al., "Microbiomes Are Integral to Conservation of Parasitic Arthropods," *Biological Conservation* 250 (2020): 108695.
27. Liz Ecker (Slausson) and Gary Paul Nabhan, "Thornber's Fishhook Cacti (*Mammillaria thornberi*) and Mycorrhizae," *Agave*, 1988.
28. Luke, "The Arizona-Sonora Desert Museum."
29. Hope Jahren, *Lab Girl* (New York: Knopf, 2016).
30. Laura Fravel, "Critics Question Zoos' Commitment to Conservation," *National Geographic*, November 13, 2003, https://www.nationalgeographic.com/animals/article/news-zoo-commitment-conservation-critic.

7

THIS IS A ZOO?

Reflections on a Wilder Zoo by Visitors to the Arizona-Sonora Desert Museum

DEBRA COLODNER, CRAIG IVANYI, AND CASSANDRA LYON

Only to the white man was nature a wilderness.

—CHIEF LUTHER STANDING BEAR, OGLALA LAKOTA SIOUX NATION, LAND OF THE SPOTTED EAGLE, 1933

AN INTRODUCTION TO THE ARIZONA-SONORA DESERT MUSEUM

The Arizona-Sonora Desert Museum immerses visitors in stories of the Sonoran Desert—its geology, climate, plants, animals, and people—and weaves these threads together into a multilayered natural history of this fascinating place. The Museum's carefully managed aesthetic recreates a series of Sonoran Desert–region habitats, from mountain forests to desert valleys, emphasizing biotic communities over singular animals and maximizing use of the viewshed in all directions (figure 7.1). As mentioned earlier in this volume, in November 2021, the Museum welcomed the authors of this book for a two-day workshop to ponder the symbiotic relationship between zoos and notions of the wild, in a place where the boundaries between zoo and wild are easily confused. Ben and Harry (the workshop conveners and volume editors) proposed that a visit

FIGURE 7.1. The Arizona-Sonora Desert Museum's location in the Tucson Mountains provides excellent views of the landscape of the Sonoran Desert from many places on the Museum's grounds.

Source: Photo courtesy of the Arizona-Sonora Desert Museum.

to the Museum might catalyze new ideas about the future roles of zoos in teaching about and conserving "the wild," and many of those ideas are now reflected in this book.

The Desert Museum was founded by two migrants to the region (from New York and Ohio) who saw the need for an educational facility to teach fellow newcomers about the Sonoran Desert's beauty, fragility, and value. Starting in the 1940s with the widespread availability of air conditioning, people flocked to Arizona from all over the United States, blading the desert to make way for a sprawling city. The Museum's founders believed that its mission of "inspiring people to live in harmony with the natural world by fostering love, appreciation, and understanding of the Sonoran Desert" was best served by interpreting nature holistically, underscoring the interconnectedness of species, habitats, biotic communities, biomes, climate, and geology. From the beginning, native animals were displayed and celebrated as ambassadors for regional conservation.

Seven decades later, the Museum has evolved into a world-renowned botanical garden, natural and cultural history museum, conservation organization, art education center, and accredited zoo. As noted by Timothy Luke in his classic and still provocative paper "The Arizona-Sonora Desert Museum—Imagineering Southwestern Environments as Hyperreality," the ASDM's imagineers created an idealized desert, with more biodiversity in its twenty-one interpreted acres than occurs in an equal area of the natural Sonoran Desert.[1] Visitors can observe many things they are unlikely to see in nearby Saguaro National Park, and the diversity of animals is the main attraction for most.[2] The Museum has become a major tourist destination, consistently attracting the audience required for both mission success and financial viability (figure 7.2).

FIGURE 7.2. About 400,000 people visit the Arizona-Sonora Desert Museum each year.

Source: Photo courtesy of the Arizona-Sonora Desert Museum.

The Desert Museum is classified as a zoo; it has nearly two hundred species of animals under human care and presents them to the public for a fee (figure 7.3). Like other accredited zoos, it holds animals for both direct conservation (for example, breeding or holding threatened and endangered species for future release) and conservation education. The Museum also serves as a sanctuary; many of the animals are "rescued" from the wild and cannot be safely returned because they have little chance of survival. Some animals spend their whole life at the Museum; others only reside there for a short time before transfer to other zoos or conservation programs. A small percentage are part of captive breeding programs, reducing collecting pressure on wild populations or creating the genetic stock necessary to augment or even reestablish wild populations. These are a few of the ways the ASDM fits the mold typical of accredited zoos.

FIGURE 7.3. Some of the animal exhibits, especially the popular Desert Loop Trail, blend seamlessly into the landscape.

Source: Photo courtesy of the Arizona-Sonora Desert Museum.

During the winter and spring of 2018, the Desert Museum joined six zoos and aquariums as sites for the third phase of the Why Zoos and Aquariums Matter study.[3] This round of research focused on STEM learning in zoos and aquariums within the informal science-learning ecosystem. Some of the Desert Museum visitor responses surprised us, especially the survey participants who did not consider the Museum to be a zoo. One visitor explained that he was generally critical of zoos but did not put the Desert Museum in that category. Another respondent looked quizzically at the survey collector after hearing that this was a study about zoos and asked, "This is a zoo?"

This piqued our curiosity. Why do some visitors not categorize the Desert Museum as a zoo? What does "zoo" mean to them? Which features of the Museum were most and least zooish in visitors' minds? Is it possible to be "just a little zooish," or is this a binary category? How might this relate to a wider perception of the Museum and its relevance to the "wild" or to "wildness"? In addition to the opinions of scholars and experts we would hear from at the Wilder Kingdom workshop, we wanted to understand more about how the Museum's members and visitors perceive it.

To begin to explore these questions, we developed a loosely structured survey to elicit open-ended responses. The survey invitation went out in the Museum's e-newsletter in the spring of 2021. The ninety-six respondents were a sample of people who take the time to read the Museum's e-news—therefore, more of an "inner circle" than a broad audience. Eighty percent of respondents were members, 12 percent were former members, and 8 percent were nonmembers. Sixty percent had visited more than once in the last year, and 97 percent of those who answered the question had visited other local nature-based sites (such as Saguaro National Park, the Coronado National Forest, etc.). Most were very aware of wildlife in their neighborhoods, and many were active hikers, birders, and so on. Many respondents were very familiar with the Museum, and their answers showed that they have thought deeply about it. Their answers were coded for general themes and categories.

WHAT DID PEOPLE SAY?

Categorizing the Desert Museum Experience

First, we wanted to know what people most enjoyed about their experiences in nature and how this compared to their experiences with zoos in general versus the Desert Museum in particular. The most enjoyable nature experience for respondents was observing animals, followed by learning, and then the sense of calm, escape, or healing nature provides. Not surprisingly, when asked what they liked most about their visits to zoos in general, animals topped their lists (in the words of one respondent, "The animals, duh!").

When asked what they liked most about their experiences at the Desert Museum, respondents brought up the botanical collections, naturalistic enclosures, and the continuity between exhibits and the surrounding landscape (figure 7.4). Interestingly, people focused on these features as much as the animals. Plants were called out in about 40 percent of the responses about the Desert Museum, compared to about 10 percent of the responses about other zoos. Naturalistic exhibits were highlighted 40 percent of the time for the Desert Museum but only in about 20 percent of the responses about other zoos (table 7.1).

We asked visitors whether they consider the Desert Museum to be a zoo and also what it is about the Museum that makes it similar to or different from a zoo in their view. Fifty-seven percent of respondents said they consider the Museum to be a zoo, 25 percent said they weren't sure, and 18 percent said they did not think it was a

FIGURE 7.4. The botanical gardens are a major feature of the Arizona-Sonora Desert Museum, equal in stature to the animal exhibits in the eyes of many visitors.

Source: Photo courtesy of the Arizona-Sonora Desert Museum.

Table 7.1. What do you like most about . . .

Experiences	Desert Museum	Other zoos
Seeing animals	26	53
Natural environments / habitats	32	18
Seeing plants / nonanimal exhibits	26	7

Number of mentions of these experiences in open-ended responses.

zoo (figure 7.5). Whether or not they thought of the Desert Museum as a zoo, respondents had similar ideas about how the Desert Museum is like or unlike other zoos. They explained that what makes the Museum different from a zoo was the feeling that it fits seamlessly with the local landscape and climate. About 20 percent mentioned the botanical gardens specifically, and 20 percent talked about zoo features they considered unique to the Museum, like the natural history and art exhibits. Another 10 percent mentioned the Museum's conservation and education work. Many respondents gave multiple reasons. A typical answer was: "It's a zoo in that it has animals as exhibits. It's not a zoo in that the ecosystem(s) is/are on display as much as the animals themselves. I really like the context."

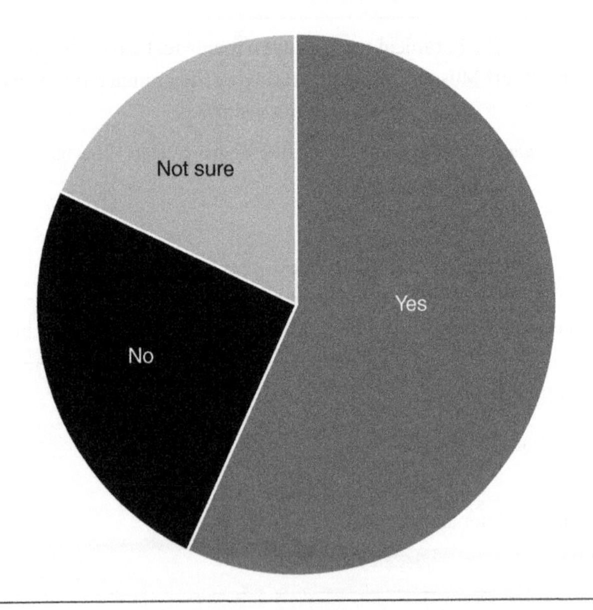

FIGURE 7.5. Do you consider the Desert Museum to be a zoo?

Total: 77 respondents.

In assessing the zooness of the Desert Museum, respondents seemed more focused on their own experiences and expectations than on the animals' experiences.[4] In contrasting the Museum with other zoos, the botanical/landscape design and regional coherence were emphasized more often than perceived differences or similarities in animal welfare. Only about 20 percent of respondents mentioned animal well-being as a factor in their judgment about the zooness of the Museum. Some felt it was important that the Museum exhibited only species native to the region and that the animals were in exhibits that closely mimic their natural habitats. A few referenced the fact that the Museum displays mostly rescued individuals, and some expressed their belief that its exhibits give animals more space than other zoos.

The Concept of "Wild"

The ninety-six people who responded to this survey expressed an impressive range of feelings about the wild. Their responses exhibited the complex and sometimes contradictory feelings about nature and wilderness in the dominant Western culture.

We asked participants about their recent experiences in nature and what they enjoyed about them. Many respondents identified nature as a part of their daily lives and not just "out there" in special, protected, or idealized places. When asked about recent experiences in nature, about 40 percent of them talked about experiences at home or around their neighborhoods. Forty percent also mentioned commonly seen urban wildlife, especially birds, as part of their recent nature experiences. Some clearly see themselves as part of nature: "[I enjoy] being quiet, able to observe how my landscape design and maintenance invites in and supports the local web of life."

Others, however, see themselves as separate: "What don't I love [about nature] . . . It's humans that are the problem."

We then asked respondents to further reflect on the idea of the "wild." Specifically, we wanted to know what they associated with the word "wild" and what in particular makes a place wild to them. Eighty percent of respondents defined "wild" as the absence of humans, human control, built structures, etc. Many mentioned the absence of barriers and fences, a pattern no doubt influenced by the context of the study. Most people had positive feelings about wild places as somewhere they like to spend time. A significant minority felt that wild places were not safe and that they would prefer to be in the "managed wild" of a place like the Desert Museum or another zoo. Several respondents openly struggled with the idea that their presence in nature decreases the wildness of any place.

Continuing, we asked whether they have ever had a wild experience at a zoo, and if so, what happened? Most people (60 percent) said no. For those who said yes, about half of them mentioned observing zoo animals. Close-up encounters or experiences where zoo animals appear to have the most freedom were perceived as more wild. About a quarter of our respondents described seeing wild (noncollection) animals at a zoo, including birds, insects, and reptiles. They also mentioned drive-through (safari-type) zoos and animal-encounter experiences. Very few mentioned natural animal behaviors as a sign of wildness. This range of experiences further illuminates the complex ways people conceive of "wild."

Many respondents had a nuanced understanding of "wild" places. One replied, "Somewhere that animals and plants can interact without major human interference." Another answered, "There is nowhere on earth where humans haven't had some kind of impact—even wilderness areas." Others had a more rigid definition, generally involving the lack of humans: "Untouched, where species live in nature in their natural state." The inherent contradictions were sometimes evident within a single response: "A place is wild when it is left to its own ends. A place can still be wild even if humankind has provided a structure for it to grow from."

It seemed that, for some, "wild" meant absence of humans and human control. Many recognized the paradox of their presence by qualifying the experiences as "more wild" with "fewer people," highlighting that absolutes were impossible. Others focused on the feeling of connecting with another species and a "wilder" part of themselves. Still others focused on the safety/danger aspects of "wild." Their answers reflected some of the central cultural myths about wilderness described by the historian William Cronon in his much-discussed essay "The Trouble with Wilderness":

> Wilderness is not quite what it seems. Far from being the one place on earth that stands apart from humanity, it is quite profoundly a human creation—indeed, the creation of very particular human cultures at very particular moments in human history. It is not a pristine sanctuary where the last remnant of an untouched, endangered, but still transcendent nature can for at least a little while longer be encountered without the contaminating taint of civilization. Instead, it is a product of that civilization. . . . Wilderness hides its unnaturalness behind a mask that is all the more beguiling because it seems so natural. As we gaze into the mirror it holds up for us, we too easily imagine that what we behold is Nature, when in fact we see the reflection of our own unexamined longings and desires.[5]

In the context of zoos, it is interesting that some of the most scripted and managed experiences (such as animal demonstrations or feedings), those requiring the most training and bending of animal behavior to human will, are the experiences many visitors identified as feeling most "wild." Those up-close encounters, where people got to feel a connection to a nonhuman, triggered a sense of wildness and perhaps stirred those "unexamined longings and desires."

We were curious whether our respondents viewed spending time in the wilderness in a positive light. So we asked them whether, when they think about a wild place, it's somewhere they want to be. Most (more than 80 percent) respondents did want to be somewhere wild. They mentioned getting away from people and the pressures of society, being themselves, enjoying the randomness, and learning more about the environment. Those who did not want to be in a wild place most often cited safety concerns. Here as well, several people mentioned the seeming contradiction in their presence making a place not wild, for example, "If I'm there then I'm mucking about somewhere that is less wild because of my presence." Several mentioned the internal nature of wilderness: "A natural setting with few humans. A place of solitude. It is a state of being not a specific place." Many also implied their safety concerns with "Yes, but" answers: "Yes, but in a controlled environment. Don't want to be in a jungle with wild animals."

Do our respondents consider the Desert Museum to be a wild place? What would make it seem wilder? These questions focused on eliciting ideas about how varied definitions of the wild related to experiences in zoos more specifically. Half of the respondents felt that the Museum is not a wild place, with the remainder evenly split between "yes" and "not sure." Some typical answers were: "No—As long as the animals are in cages, the Museum can't be wild" or "Yes—The environment is in a natural state. Huge lizards and snakes have the run of the place."[6]

In terms of *making it seem wilder*, the most common responses were variations on the theme that this is not possible and, importantly, that there was no need to try. Others focused on the size and design of enclosures, assuming that larger enclosures, which more closely simulate natural habitats, would make the experience wilder (and better) for the animals. Many focused on the botanical experience for human visitors, suggesting (and preferring) a more haphazard and less manicured feel. Several mentioned putting humans in the more contained space and drive-through zoos, as well as having fewer captive animals. About 10 percent felt that being able to better see and interact with the animals would make it seem wilder. These responses again demonstrate the conflict between expectations and reality; close animal-human interactions require the most human intervention and control, yet the feeling of seeing/touching and being seen/touched by an "other" can evoke a deep sense of the wild (figure 7.6).

Overall, respondents emphasized enjoying the natural feel of the Desert Museum, while recognizing it is curated by people, for

FIGURE 7.6. A great-horned owl performs in the museum's Raptor Free Flight program. Raptor Free Flight is an example of a program where the animals appear to be wilder but in fact are highly trained and managed.

Source: Photo courtesy of the Arizona-Sonora Desert Museum.

people. They appear to place it on a continuum between human-dominated spaces and the "wilderness" (culturally defined) and find it closer to the wild than most zoos. In thinking about making the Museum "more wild," this sample of Museum visitors did not anticipate some of the proposals being debated among zoo and animal professionals, aimed at allowing zoo animals a fuller range of their natural behaviors, such as hunting prey or more opportunities for reproductive sex. These respondents did not stray from the peaceful, harmonious, and sanitized view of nature presented in American zoos.

CONCLUSION

As they write in their framing essay, Ben and Harry posed a series of questions to the authors in this book. In the spirit of increasing the positive impact of zoos and hypothesizing that the Arizona-Sonora Desert Museum might be a model for other institutions, we focused our inquiry on whether people view it as a zoo, whether zoos can be wild, and, if so, what a wilder zoo might look like.

Based on our survey results, as well as anecdotal information, it's clear most people view the Desert Museum as a zoo but as a unique type of zoo. Survey responses showed how visitors recognize and appreciate the Museum's regional context, holistic approach, and immersive feel. The majority of respondents showed an affinity for wild experiences and did not view zoos as wild but described the Desert Museum as *closer* to the wild than most zoos. They also provided feedback about how the Museum could be or at least feel wilder. Respondents pointed to increased enclosure size and making the Museum's grounds feel less groomed, along with reducing the number of live animals it interprets.

Interestingly, our survey data align with results from the Why Zoos and Aquariums Matter study, which found that the area of least trust of zoos and aquariums had to do with enclosure size compared to the perceived needs of the animals.[7] In particular, romantic notions of "the wild" as a vast unbounded space can mislead people to wrongly intuit that animals need a lot of space for their well-being. However, studies have repeatedly shown that the welfare of different species depends on different combinations of physical and mental stimulation that do not necessarily rely on space.

One contradictory result in our survey is that a significant minority of respondents suggested that increasing the number of animals and encounters with them (which would presumably give animals less space) would enhance a feeling of wildness. This is a conundrum, in that seeing many animals (especially up close) is not what people typically experience in the wild, where the frequency and proximity of animal encounters is low. Moreover, close-up experiences with zoo animals often require the highest level of staff interaction and manipulation of animal behavior to ensure visitor and animal safety.

It is apparent from survey respondents that the Museum could take steps to be or at least *feel* wilder. However, just because it is possible to make a zoo wilder doesn't necessarily mean we should. Luke posits that creating hyperreal experiences like the Desert Museum distorts visitor perception of the wild and draws more people to visit, establish residence in, develop, and consume the Sonoran Desert. This echoes the general debate about ecotourism—the conflicting push and pull between asking people to keep a respectful distance and inviting people in—and it is always important to acknowledge the tradeoffs of these endeavors. In most cases, however, development is driven by forces well outside the purview of zoos and similar institutions. Their existence does not encourage or dissuade it but may shape development into more nature-friendly dimensions. Some long-term Tucson residents and community leaders credit the Desert Museum with doing just that.[8]

Most of the visitors' responses speak to a holistic view of nature. These visitors find "wildness" or "nature" wherever they see or interact with animals or plants. In general, respondents viewed the wild as less tamed, controlled, and touched by humans. They showed awareness that even the wildest zoo is an illusion, not the true wild, if there is such a place. The zoo translates and imitates an idealized wild. At the same time, the "wild" is being transformed into the "zoo"; in one form or another, human hands touch every place on the globe, restricting nature in enclosures of different scale and management scope. Does the zoo's portrayal of idealized "harmony with the natural world" encourage us to walk more softly or run more carelessly through the rest of nature as we shape it to suit our needs? Can a wilder zoo increase the compassion we show to the rest of nature as we pursue our ends? Zoos and the scholars who study them will need to continue to ask our visitors to help us answer these questions.

The Desert Museum is a special place, first envisioned and then established by people with a unique mission. This institution was built with the surrounding environment front of mind. Unlike many zoos, which center visitor experiences with individual animals to encourage a sense of connection with global nature, the Museum centers the Sonoran Desert ecosystem. As discussed by Susan Clayton in this volume, both approaches have value. Developing a shared identity with individual animals appears to encourage development of a conservation ethic, as does a deeper understanding of interconnected ecosystems. Where other zoos have to improvise, allude, and imagine a connected habitat, the Desert Museum can simply invite their visitors to look around, look more closely, and see the multitude of relationships among themselves and wild animals, plants, and the larger Sonoran Desert ecosystem. It is these relationships that Gary Nabhan, Alison Deming, and others in this volume remind us should be an increasing focus of zoo conservation and education efforts.

As the zoo community continues to be invoked in a myriad of dichotomous discussions ranging from whether zoos are "good" or "bad" to whether they "support conservation" or "harm conservation," our survey results emphasize the importance of increased nuance. One increasingly poignant piece of the debates around zoos relates to their capacity for wildness. To some, a zoo can never be wild, given its human-focused aims and constructed nature. To others, a zoo is the only place they will be confronted with "wild" animals or ecosystems while still feeling safe. The Desert Museum is unique: it refrains from importing the idea of wild from other ecosystems and instead highlights the wild available in visitors' own backyards. The Museum's legacy of dedication to the surrounding desert provides an example for other zoos of how focusing on local ecosystems can provide abundant opportunities for visitors to discover the delights and constraints of the web of relationships in which they live.

NOTES

1. Timothy W. Luke, "The Arizona-Sonora Desert Museum: Imagineering Southwestern Environments as Hyperreality," *Organization and Environment* 10 (1997): 148–63.
2. Intrinsic Impact Surveys, Arizona-Sonora Desert Museum 2017–2019.
3. Rupanwita Gupta et al., "Public Perceptions of the STEM Learning Ecology—Perspectives from a National Sample in the US," *International Journal of Science Education, Part B*, 10, no. 2 (2020): 112–26, https://doi.org/10.1080/21548455.2020.1719291.
4. As Nigel Rothfels pointed out in the workshop for this book, zoos are built for people by people. Additionally, as other contributors to this volume write, humans are inextricably linked to our understanding of the "wild" and "wilderness" and perhaps should be included more obviously in discussions of nature in zoos.

5. William Cronon, "The Trouble with Wilderness; or, Getting Back to the Wrong Nature," in *Uncommon Ground: Toward Reinventing Nature*, ed. William Cronon (New York: Norton, 1995), 69–90.
6. The huge lizards referred to in this comment are spiny-tailed iguanas, an introduced species from Sonora, Mexico. Although they run free on the museum grounds and the museum does not manage them, their presence is the result of human interference decades ago. Ironically, this example of a "wild" animal is not in its native range. They depend on the enhanced resources at the museum and have not established themselves off of the museum's grounds.
7. Shelley J. Rank, et al., "Understanding Organizational Trust of Zoos and Aquariums," in *Understanding the Role of Trust and Credibility in Science Communication*, ed. Kathleen P. Hunt (Ames: Iowa State University Summer Symposium on Science Communication, 2018), https://doi.org/10.31274/sciencecommunication-181114-16.
8. For example, Dale Turner, Nature Conservancy, personal communication; William Shaw, University of Arizona, personal communication.

8

EVOLUTION TO THE RESCUE

Natural Selection Can Help Captive Populations Adapt to a Changing World

JONATHAN B. LOSOS

From the origin of coral atolls, to how plants colonize distant islands, to the formation of soil, Charles Darwin was remarkably prescient in his scientific ideas. Not to mention his greatest idea of them all, the theory of evolution by natural selection. Yet Darwin got two things wrong.[1] One was the mechanism of heredity. A century before Watson and Crick, Darwin's ideas of pangenesis were completely mistaken.

The other misguided idea was Darwin's view on the pace of evolution. Darwin famously believed that evolution occurred at a glacial speed, much too slowly to be observed except over the span of many millennia or longer. "We see nothing of these slow changes in progress, until the hand of time has marked the long lapse of ages," he wrote in chapter 4 of *On the Origin of Species*.

In his defense, there were no data available at the time to inform Darwin's views; no one was studying evolution, much less the rate at which it occurred. Consequently, Darwin leaned on Lyell's theories of gradual geological change, as well as Victorian views of the appropriate pace of transformation in society—slow (remember, this was the era of the French and industrial revolutions)!

Darwin's views on the pace of evolution held sway for more than a century, but we now know that he was mistaken. Evidence from the peppered moth; the evolution of insecticide, herbicide, and antibiotic resistance; field studies in the Galápagos and many other places; and even field experiments have now made crystal clear that when the environment changes—when directional natural selection is strong—then evolutionary change can occur rapidly, so fast that it can sometimes be observed as it unfolds over the span of a few years.

This realization has profound implications for our conception of wildness and how we should conserve it. Historically, the conservation movement has had as its goal the preservation or restoration of the state of nature relative to a baseline state that occurred before humans came along and messed things up. In principle, this is easy enough. Simply decide on the appropriate baseline—be it the evolution of *Homo sapiens*, the arrival of humans to a particular place, or European colonization—and use that to measure conservation success.

But if the world itself is constantly changing and species adapting to these changes, then evolutionary processes must be incorporated into conservation planning. We can't just restore the wild to the world as it was, because that world no longer exists. And this perspective applies as much to zoos as to any other conservation practitioner.

From its inception, the field of conservation biology included evolutionary considerations. Indeed, Frankel and Soulé's *Conservation and Evolution* was one of the foundational books in the nascent field.[2]

In the early days, however, the focus was on the amount of genetic variation retained within endangered populations.[3] Because of their small size, such populations are at risk of several genetic maladies. Small populations tend to lose genetic variation; this loss is undesirable because individuals lacking genetic variation tend to be less fit. Moreover, genetic drift—random changes in the frequency of different genetic variants—tends to occur at a rate inversely related to the size of the population. The risk is not only that some variation may be lost, but also that deleterious variants—those that cause disease or other problems—might become predominant because of random fluctuations. Finally, over the long term, genetic variation is the key to evolutionary success. Without variation, a population does not have the raw material available for natural selection to use in sculpting new adaptive solutions.

For those reasons, early conservation biologists, particularly those managing populations in zoos or elsewhere, focused on means to minimize the loss of genetic variation in populations, as well as to minimize the breeding of closely related individuals, which could lead to inbreeding problems. These concerns led to today's sophisticated methods for managing studbooks of captive populations.

At the same time, a second evolutionary issue was recognized. Zoo environments differ from those a species experiences in the wild. The risk exists that over generations, populations may adapt to their zoo environments in ways that would not serve them well if and when reintroduction back into the native environment occurs. To cite one example, individual animals differ in how nervous they are, how flighty and responsive they are to external stimuli. It's easy to imagine that a nervous antelope might respond to the backfiring of a zoo cart by fleeing in the opposite direction, directly into a wall or fence, breaking a leg or its neck. In this way, natural selection would favor zoo animals that are less responsive to such stimuli. Some have considered this process of becoming less reactive to external stimuli to be very similar to what occurred during the domestication of many species (see Wynne and Molinario's chapter in this volume)—zoos may unintentionally be removing the "wild" from their residents. Such evolutionary change would be adaptive in the zoo context, but we can well imagine that if individuals from such populations were subsequently reintroduced to their natural environment, these placid, unresponsive animals might not fare well in the presence of predators.

Although this example is extreme, selection could occur in more subtle ways. Maybe nervousness correlates with ability to raise babies, or perhaps the difference in day length between the natural environment and the zoo's location affects some individuals more than others. In many ways, natural selection could lead to adaptation to conditions in captivity, but detecting it—and thus realizing that it is occurring—would be more difficult.

Let me cite one relevant example, though not from a zoo. The rapidity by which natural selection can lead to the loss of antipredator behavior can be surprising. We know this because of a conservation project taken to save the northern quoll, an adorable spotted marsupial from Australia that bears some resemblance to a cat, hence its other name, the "marsupial cat." Unfortunately, the northern quoll has fallen on hard times. In an effort to establish a safety population, conservationists in 2003 moved forty-five quolls to a predator-free island off the northern coast of Australia. The quolls thrived in their new paradise, so much so that thirteen years later, conservationists relocated some individuals back to the mainland in an attempt to reestablish them in an area where they no longer occurred. Unfortunately, the reintroduction failed because the quolls quickly were killed by dingoes. Subsequent study revealed that in a mere thirteen generations (male northern quolls normally live only a year, females slightly longer) on the island, the marsupials had completely lost the ability to detect and react to the scent of dingoes, a remarkable example of rapid evolutionary loss of antipredator behavior.[4] Presumably, island individuals that were less cautious fared better—perhaps they spent more time looking for food—and thus wariness disappeared from the population. The same could easily happen in a zoo population.

In an example somewhat closer to zoos, fishery managers have shown that salmon adapt rapidly to being raised in captivity in hatcheries, but this adaptation comes at the cost of poorer survival when released back into nature. The cause of this adaptive tradeoff is not yet understood.[5]

The solution in zoos to this problem is the same as with maintaining genetic variation: make sure that individuals contribute equally to the next generation, thus minimizing the loss of variation from one generation to the next. Of course, this approach needs to be balanced by not allowing individuals with obviously deleterious traits to breed—removing such variation from the population is important.

To summarize, the traditional approach to evolution vis-à-vis management of captive populations has been to try to keep populations the same, unchanged and undomesticated. When the time comes for reintroduction back into nature, such populations will thus be just as well adapted as the day they were brought into captivity.

But there's a problem. Nature is not static. Environments are always in flux, particularly in these fraught times with the climate changing and humans disrupting the environment in so many ways. We may consider an unchanged population the ideal of what we are trying to preserve—wild, unspoiled by its time under human care—but that population may no longer be well adapted to the world into which we'd like to reintroduce it. And that's where the positive aspect of natural selection comes into play. Instead of weeding out new variations to maintain the status quo, managers can take advantage of the power of natural selection to help prepare populations to survive and adapt to these new conditions. The wild is constantly changing, and to be relevant, zoo populations need to evolve in the same way.

This idea will be the focus of the remainder of this chapter.

I'll start with a hypothetical example. There's great concern that as the planet warms, many lizard species will go extinct. Already, many populations of many species have diminished or disappeared entirely, with higher temperatures the most likely cause.[6]

We herpetologists have a good understanding of how to determine the temperature at which a lizard functions the best (remember, lizards do not maintain a constant body temperature like mammals do; rather, their body temperature is related to the external environment). By running a lizard on a racetrack at different body temperatures (or measuring some other aspect of physical performance), we can determine at which body temperature the lizard performs best.

In most cases, lizard populations are well adapted to their local conditions, with maximal performance capability corresponding to temperatures they most frequently experience. Nonetheless, as occurs for many traits, variation in this optimal temperature for performance will exist from one individual to the next: one might run most quickly at eighty-five degrees Fahrenheit, another at eight-six, yet a third at eighty-seven.[7]

Now, suppose you are breeding an endangered species that you would like to eventually return to the desert from which it came. And suppose, further, that the average optimal body temperature for the population is eighty-five, which corresponds well to what the typical temperature was thirty years ago, when the population was brought into captivity. However, in that time, temperatures at that site have increased on average by two degrees. If we put the lizards back as they are, the body temperatures they experience may end up being higher than their optimal temperature; in fact, they might dangerously overheat.[8]

Here's what we do. We measure the optimal performance temperatures of each lizard in our breeding colony and then only allow the ones with the highest optimal temperatures to breed. This is called "artificial selection," the same process by which our agricultural plants and animals were shaped. Experiments in the laboratory on fruit flies and many other organisms show that artificial selection is almost always successful in leading to evolutionary change in the desired selection. So, by selecting on higher optimal body temperatures, we should be able to produce lizards better adapted to new conditions in their natal homeland, ready for reintroduction.

Now, let's take a real-world example based on remarkable work taking place in central Australia. Although the work is being conducted in very large, outdoor areas, the same principles can be applied in zoos and certainly in the larger open areas maintained by some zoos in satellite facilities.

In many parts of the world, introduced predators have wreaked havoc on native species. If the goal is to reintroduce those prey species, there are only two options: either get rid of the introduced predators or find a way for the prey species to coexist with the predators.

The go-to method for conservationists has been predator removal by gun, poison, or any means available. Sometimes this works, but many introduced species are difficult to eradicate. This is certainly the case in Australia, where introduced domestic cats and red foxes have had cataclysmic effects on small mammal communities. The situation is so dire that for threatened species like bandicoots and hare-wallabies "most of the Australian land area is now effectively uninhabitable" because of "predation by just two introduced species," according to a team of leading Australian conservationists.[9]

Despite intensive predator control methods, the wily predators are still abundant. Many conservationists feel that the effort is a lost cause. Some other means is needed to get native fauna back into the wild.

Australian researchers at Arid Recovery, a research center in South Australia, had a bold idea to take the evolution approach one step further than the artificial selection I outlined for lizard thermal biology. Instead of posing the selective pressures themselves, they proposed to let nature take charge.

Arid Recovery started out as a conservation center, erecting cat- and fox-proof fences to enclose large areas in which threatened species of mammals could thrive in the absence of predation. And thrive they did. But the ultimate goal was not to maintain these species behind fences but to give them the ability to survive in the wild, outside the fences and in the presence of predators.[10]

The researchers decided to take an unheard of, radical, approach. Why not conduct experiments in nature, releasing naïve prey in the presence of unrestrained cats? The hope was that the prey, either through learning or evolving, would find a way to survive in the presence of cats. It was a big gamble—these were highly threatened species—but Arid Recovery had healthy populations, enough to use some individuals as guinea pigs, and several large enclosures designed for just this sort of experiment.

The trick, the researchers reasoned, would be to put enough cats in the enclosure to promote the development of appropriate behaviors in the prey, but not so many that the prey populations would be wiped out before they had a chance to learn or evolve. The gamble, of course, was that such a sweet spot actually exists, that there is a density of cats low enough that the prey species can survive in their presence.

To find out, the researchers introduced into a ten-square-mile enclosure 352 boodies (a small type of kangaroo about the size of a rabbit, otherwise known as the burrowing bettong) and forty-seven bilbies (chinchilla-esque small marsupials with powerful limbs, an elongated snout, and donkey-like pink ears, seemingly twice as large as they should be). A year later, five feral cats from outside the fence were captured, neutered, and released inside the enclosure, joining a sixth cat that had slipped in on its own. The resulting density was equivalent to the lower end of the range of cat densities in Australia. As a control, boodies and bilbies were added to several of the predator-free exclosures.

The experiment had two main questions. First, would the boodie and bilby populations be able to survive, even thrive, in the presence of low densities of cats? Second, would the exposure to cats lead the populations to up their antipredator game?

Eighteen months later, the results were clear: both bilbies and boodies can survive in the presence of felines.[11] The boodie population had more than doubled; 65 percent of the adult females had young in their pouches. The size of the bilby population couldn't be estimated from trapping for the simple reason that the traps were so full of boodies that few bilbies were caught. But camera traps recorded bilbies ten times more often than they had at the start of the project, and the number of bilby tracks counted in the sand more than tripled.

So the populations were thriving, but how were they doing it? Were the boodies and bilbies becoming better at living in the lion's den?

To find out, the researchers measured the boodies' wariness over the course of the experiment, comparing animals living with cats to those in the predator-free exclosure (which served as the control). Within eighteen months, differences were already apparent.[12]

This increase in wariness could come about in two ways. One possibility is that individual boodies and bilbies changed their behavior once they began living in the company of cats. Because some of the boodies were radio-collared and could be tested repeatedly, we know this occurred, though not how. Did they learn from personal experience after surviving an attack or observing an attack on others? Or did a few individuals figure it out and then somehow pass that knowledge on to others? Boodies live in groups, which would facilitate exchanging tips, trading life hacks as it were. Or maybe mothers pass on their knowledge to their young? We don't know how it happened.

In addition to individuals learning how to survive in the presence of cats, there's another way that the populations could have gotten better at existing with felines, a radical idea that the researchers championed. What if the cats were agents of natural selection, preying on the individuals least capable of avoiding predation? Survival of the wariest or fastest, perhaps. And if those traits have a genetic basis, then the population should evolve from one generation to the next, becoming increasingly more capable to withstand cat predation.

The hope, in other words, was to harness the power of Darwin to produce Australian prey species capable of living with cats. In theory, it's possible—we know that evolution can occur rapidly when natural selection is strong. To examine this possibility further, the researchers compared the anatomy of the animals in the predator and control enclosures. One piece of evidence was tantalizing: after four years with cats, the average hindleg length of male boodies born in the cat enclosure was nearly 2 percent longer than the control animals. One possibility is that longer-legged boodies are faster or more agile and that natural selection had favored the evolution of this long-leggedness.[13]

The researchers tested this hypothesis by introducing boodies in yet another enclosure with cats to see if longer-legged individuals survived longer. In a trial with cats present but at low abundance, longer-legged boodies did, indeed, have a survival advantage.[14]

Though not definitive, this would seem like strong evidence that the boodies had evolved in ways to make them less susceptible to cat predation. More research is needed, but these results highlight the possibility that natural selection could pave the way for coexistence.

These trends were all very promising. But was it worth all the time and effort? Did the new tricks they'd learned and the new traits they may have evolved make the animals from the cat enclosure better able survive in the presence of cats?

There was only one way to find out: put animals from both populations into the same enclosure and let them go mano a mano. The hypothesis was straightforward: if living with cats enhanced survivability, the predator-enclosure marsupials should outlast the animals from the control paddock that had been living blissfully cat-free.

To test this idea, the researchers switched species. Forty-seven bilbies—twenty-three from the cat enclosure, twenty-four from the control—were released into another large enclosure containing ten cats.[15] All the bilbies had radio-collars attached so that the researchers could relocate them, dead or alive. One disadvantage of this approach, however, was that the batteries on the radio-transmitters only lasted forty days.

Even that short window was long enough to answer the question. Survival of the bilbies that had grown up living with cats was twice as high as that of cat-naïve marsupials, more than two-thirds of which were dead by the end of the study period. Living with real live predators is good training for life in the wild.

The use of natural selection to make threatened species better adapted to live in the wild is just in its infancy, but the possibility is clear. Scientists have been studying how natural selection operates for decades, so the tools are available. As the two examples

I presented—one hypothetical, the other real—demonstrate, it is possible to help species adapt to a changing world.

The wild is in flux, so we will need to actively and intelligently manage captive populations slated for reintroduction so that they are maximally adapted to future environments that differ in landscape, climate, and many other ways from the wild they formerly occupied. This necessity is a reminder that any functional and realistic understanding of the wild, both outside and in the zoo, requires a degree of intervention if we want to achieve broader conservation goals.

Species are going to need all the help they can get on our rapidly changing planet. And still relevant after all these years, Charles Darwin is showing the way.

NOTES

I thank Ben Minteer, Harry Greene, Craig Ivanyi, and all the participants in the Wilder Zoo discussion for the opportunity to write this chapter. The research was supported by the grant "Rethinking Zoo Biology: The Histories, Effects and Futures of Captivity," funded by the Australian Research Council (DP200103404).

1. Worrying that somehow doing so will empower critics of evolution, some of my colleagues criticize me for saying that Darwin was wrong about anything, as if he were an omniscient deity.
2. Otto Frankel and Michael E. Soulé, *Conservation and Evolution* (Cambridge: Cambridge University Press, 1981).
3. A nice review of the role of evolution in conservation is Albrecht I. Schulte-Hostedde and Gabriela F. Mastromonaco, "Integrating Evolution in the Management of Captive Zoo Populations," *Evolutionary Applications* 8 (2015): 415.
4. Chris J. Jolly, Jonathan K. Webb, and Ben L. Phillips, "The Perils of Paradise: An Endangered Species Conserved on an Island Loses

Antipredator Behaviours Within 13 Generations," *Biology Letters* 14 (2018): 20180222.

5. Mark R. Christie et al., "Genetic Adaptation to Captivity Can Occur in a Single Generation," *PNAS* 109 (2012): 238. For a general review, see Mark R. Christie, Michael J. Ford, and Michael S. Blouin, "On the Reproductive Success of Early-Generation Hatchery Fish in the Wild," *Evolutionary Applications* 7 (2014): 883.
6. The effect of global warming on lizard survival was brought to the world's attention in Barry Sinervo et al., "Erosion of Lizard Diversity by Climate Change and Altered Thermal Niches," *Science* 328 (2010): 894.
7. An example of studying the effect of body temperature on lizard sprinting performance is Raymond B. Huey and Paul E. Hertz, "Is a Jack-of-All-Temperatures a Master of None?," *Evolution* 38 (1984): 441.
8. I'm simplifying here—there are other options, such as that the lizards might seek out shade more often.
9. Sarah Legge et al., "Havens for Threatened Australian Mammals: The Contributions of Fenced Areas and Offshore Islands to the Protection of Mammal Species Susceptible to Introduced Predators," *Wildlife Research* 45 (2018): 641.
10. The best introduction to the fabulous work conducted at Arid Recovery is the center's website, which includes much information on what they do there, as well as a list of publications stemming from research there: https://aridrecovery.org.au/.
11. Katherine E. Moseby et al., "Understanding Predator Densities for Successful Co-existence of Alien Predators and Threatened Prey," *Austral Ecology* 44 (2019): 409.
12. Rebecca West et al., "Predator Exposure Improves Anti-Predator Responses in a Threatened Mammal," *Journal of Applied Ecology* 55 (2018): 147; Lisa A. Steindler et al., "Exposure to a Novel Predator Induces Visual Predator Recognition by Naïve Prey," *Behavioral Ecology and Sociobiology* 74 (2020): 102.
13. K. E. Moseby et al., "Designer Prey: Can Controlled Predation Accelerate Selection for Antipredator Traits in Naïve Populations?," *Biological Conservation* 217 (2018): 213.

14. H. L. Bannister et al., "Individual Traits Influence Survival of a Reintroduced Marsupial Only at Low Predator Densities," *Animal Conservation* 24 (2021): 904.
15. Alexandra K. Ross et al., "Reversing the Effects of Evolutionary Prey Naiveté Through Controlled Predator Exposure," *Journal of Applied Ecology* 56 (2019): 1761.

9

ZOO DOGS

CLIVE D. L. WYNNE AND HOLLY G. MOLINARO

There is hardly a tribe so barbarous, as not to have domesticated at least the dog.

—CHARLES DARWIN, *ORIGIN OF SPECIES*

Forty years ago, the San Diego Zoo Safari Park pioneered the presentation of a species of mammal that had seldom before been seen in a zoo setting. This rare animal was a domestic dog (*Canis lupus familiaris*). But Anna, a golden retriever, was not acquired on her merits as an exhibit; rather she was there to provide companionship to a cheetah (*Acinonyx jubatus*), Arusha. Arusha had been hand-reared and consequently unable to form relationships with her own kind. The experiment of pairing a retriever with a cheetah was considered a success, and over the subsequent decades dogs have been brought into cheetah enclosures by several other zoos in the United States (figure 9.1).[1]

In this chapter, we explore what the absence of dogs from zoos says about dogs, zoos, and tameness as a contrast to wildness. In so doing, we spotlight three misunderstandings about dogs. Exploring these confusions enables a closer look at the wild/tame boundary and shows that far from being a clear-cut distinction, this is

FIGURE 9.1. A cheetah at the San Diego Zoo with a dog as companion.

Source: Photo by Zach Tirell. https://www.flickr.com/photos/tirrell/58268769/in/faves-155419065@N04/Attribution-ShareAlike 2.0Generic.

actually a littoral zone in which questions about the interplay of biology and human culture are richly interfertile.

Revealingly, dogs are not usually found in zoos, at least not as exhibit animals—though there have been occasional exceptions. A zoo in Luohe, Henan Province, China, tried to pass off a Tibetan mastiff as an African lion, but the dog blew its cover by barking.[2] Dogs have occasionally played other roles in the zoo world. In the eighteenth century, the public were admitted to the menagerie at the Tower of London on the payment of three ha'pennies or provision of a cat or dog to be fed to the lions.[3] In the late nineteenth century, the famed zoo entrepreneur Carl Hagenbeck also featured dogs in his Hamburg, Germany, zoo, even going so far as to have them photographed next to lions for guests' amusement.[4] Dogs were fed to snakes in zoos in Florida as recently as 1996,[5] and the feeding of dogs to exhibit animals may still occur in Chinese zoos.[6] Recently, some U.S. zoos have offered dogs as "ambassador" animals. Ambassadors are usually particularly tame and handleable individuals from among exhibit animals that can be brought out of enclosures and presented to the public for closer inspection. With growing liability concerns, this practice has waned; however, at least one zoo, in Santa Barbara, California, has introduced a dog in this role. According to the zoo's vice president of animal care and health, this tame domesticated individual's role is to "connect with Zoo guests so they can understand and care about all animals, especially those in the wild."[7]

These minor exceptions notwithstanding, the absence of dogs from zoos is not hard to understand. Dogs do not fulfill any of the criteria that zoos use to select their exhibits: They are not wild, exotic, or endangered. Instead, they are the opposite of all those things: They are domesticated, familiar, and numerous.[8] Other related canids may be regulars of the modern zoo, including various other subspecies of wolves (grey wolves, Mexican wolves, Arab wolves, etc.; *C. l.* subsp.)[9] and other Canidae (such as the African wild dog, a.k.a. African painted dog, *Lycaon pictus*). The only animal routinely found in zoos sometimes listed as a member of the subspecies *C. l. l.* is the dingo. But the dingo is exotic, even to urban Australians, and some authors list the dingo as a distinct species of *C.* (*C. dingo*)[10] rather than a subspecies of *C. l.* Certainly the dingo is quite behaviorally distinct from the common domestic dog.[11]

The general absence of dogs from zoos reveals interesting caveats in the ways we think about zoos, captivity, and the wild while also flushing out three misunderstandings about dogs. We next discuss these in turn.

FIRST ERROR: DOGS ARE NOT A PRODUCT OF ARTIFICIAL SELECTION

The first mistake is to view dogs as a direct outcome of human intentions. The modern understanding of the process of domestication dates to Charles Darwin in his *On the Origin of Species*.[12] Darwin identified a process by which wild animals were turned into suitable companions for a human home by selection over many generations. To Darwin, domestication implied active human involvement in the selection and breeding of the tamest animals to be the parents of the next generation. He recognized that this did not necessarily have to be a conscious process. In addition to conscious artificial selection, Darwin also proposed "a kind of Selection, which may be called Unconscious, and which results from every one trying to possess and breed from the best individual animals."[13] Although the human action was not necessarily conscious, it was still human intentionality that shaped the domesticating species. Darwin wrote about domestication at some length in his *Origin of Species* because he wanted to exploit the everyday observation of humans selecting animals as an analogy for how "nature" herself could select forms better suited to ambient conditions. He wrote: "Why, if man can by patience select variations most useful to himself, should nature fail in selecting variations useful, under changing conditions of life, to her living products?"[14]

As Jonathan Losos points out in the preceding chapter, among his many astonishingly prescient thoughts, Charles Darwin also got a few things wrong: hardly surprising in one who was working around 150 years ago. Most contemporary researchers into the domestication of the dog view Darwin as incorrect to state that the original domestication of dogs involved human intention. Dogs are unlikely to be the outcome of artificial or even unconscious selection. Rather, particular groups of wolves adapted by natural selection to exploit the proximity of human hunter-gatherer groups.[15] At certain points during the last ice age, our ancestors found locations where the hunting was particularly rich, and they became sedentary. This probably occurred especially around the time of the last glacial maximum, about 20,000 years ago, when climatic conditions were so severe that only small pockets of the planet's surface, known as refugia, were habitable by our own and many other species.[16] Under these conditions of enforced sedentism and close proximity to other species, our ancestors would have started accumulating mounds of trash, which, then as now, attracted scavengers that foraged on the remains. These, though inedible by people, still contained nutrients other species could exploit. Among the species that colonized human middens were wolves. To this day, wolves scavenge on human trash dumps in diverse parts of the world.[17] Over time, some populations of wolves began to adapt through natural selection to be better suited to exploiting these human-generated food sources. As Losos observes, the rapidity with which natural selection can operate under these conditions would also have been a surprise to Darwin, who viewed evolution as inevitably extremely slow.

In order to maximize their effectiveness in extracting food from human communities, these wolves becoming dogs underwent natural selection, which led to several adaptations. In the domain of reproductive behavior these include reduced pair bonding, greater promiscuity, loss of seasonality of breeding, and reduction of parental investment in young.[18] This set of adaptations enables dogs to reproduce more rapidly when resources increase suddenly while reducing unneeded investment in training young to assist in bringing down live prey.

Wolves becoming dogs also underwent adaptations in their development, most notably an extension of the critical period for social imprinting.[19] Animals are not born knowing what species they belong to but rather in the first days of life use their senses to identify the kinds of beings they have been born among. They develop a template from those earliest experiences that guides their selection of social and sexual companions for the rest of their lives.[20] In wolves, as in all nondomesticated species, this process is complete in the first few weeks of life, which ensures that, short of exceptional circumstances, individuals from these species only form attachments to members of their own species. Dogs, however, show an extended period for social imprinting such that dog puppies readily form social connections with individuals from any other species they are exposed to in the first several months of life.[21] These adaptations initially helped wolves becoming dogs obtain more resources from their human hosts than could wild-type wolves. There is also a relationship here between the likely evolutionary pathway from wolf to dog and the evolution that can be observed in captive animals held in zoos and elsewhere (another point made by Losos in his chapter).

SECOND ERROR: DOGS DO NOT LIVE ONLY IN HUMAN HOMES

The second mistake is to imagine that dogs, because they are domesticated, live primarily in human homes. To say an animal is domesticated is, etymologically, to state that it is accustomed to dwell in a house.[22] Evidence for dogs in human dwellings can be traced back in the historical record at least to Arrian, an ancient Greek writing in the second century BCE.[23] Arrian advised the huntsman to let his best hound sleep on his bed with him. In the archeological record, a dog pup was found buried with a woman under a dwelling at a site in Israel dating back to the twelfth millennium BCE—which may imply that the dog was allowed inside the dwelling in life as well.[24] Ancient Egyptian artworks show dogs resting under chairs in what appear to be interior scenes.[25]

Notwithstanding the long history of the practice, the majority of the world's estimated 800 million dogs[26] do not live inside human domiciles. Precise estimates are difficult to obtain, but the majority—perhaps around 70 percent—of the world's dogs live outside human homes.[27] They are variously known as "street dogs," "village dogs," "stray," or "feral." The last two terms can be misleading. To "stray" implies that an individual has some more proper place to reside, which it has left illegitimately. Dogs living outside human homes have not typically absconded, so it makes little sense to refer to their modal form of life as if it were anomalous. "Feral" also carries connotations that the animal has left its proper place, which, again, does not seem an appropriate nomenclature for the animal's dominant lifestyle. Very few populations of dogs sustain their numbers by hunting live prey, but only a minority of dogs live inside human homes as pets.[28] Most dogs live in an intermediate state where they mostly scavenge on the residues of human food production, are often close to people and may even be identified as "belonging" to specific people, but do not live inside any human domicile.[29]

THIRD ERROR: DOMESTICATED IS NOT THE OPPOSITE OF WILD

The third mistake, which follows to some degree from the first two, is to view "domesticated" as the opposite of "wild." As noted above, domestication does not mean "found in a *domus*," and it is also not a synonym for artificial selection (at least so far as dogs are concerned). It follows that "domesticated" is not the opposite of "wild."

It will be useful here to discriminate between "domesticated" and "tame." Domesticated animals are members of a species (or subspecies) that has been selected over generations to be more tolerant of human proximity. "Domesticated" refers to phylogeny, the change in generations over time. Tame animals are individuals that have been reared by, or at least live in close proximity to, humans and are comfortable eating, sleeping, mating, and so on close by human beings. Consequently, "tame" refers to ontogeny, the change in an individual animal over the course of its lifespan. Members of domesticated (sub)species are much more likely to be tame than are individuals from wild species, but this is a matter of degree rather than an absolute boundary. Wild (nondomesticated) animals can be tamed if they are removed from their mother at a sufficiently early age and kept in human proximity. However, domesticated animals are much easier to tame. Indeed, it is so easy to tame dogs that the fact that every individual dog pup needs to be raised close to human beings if it is to accept human proximity in adulthood is not widely understood. An experimental study from the 1950s demonstrated that dog pups raised in isolation from humans until their fourteenth week of age were unable to tolerate human presence in later life.[30] Conversely, individual wild animals have often been tamed. Wolves, the ancestors of dogs, can be brought to form relationships with people if they are taken from their mothers by ten days of age and raised in continuous human contact for the first several months of life.[31] Wolves raised in this way show evidence of the same kind of attachment to human caretakers that is found in dog pups, though they do not as readily generalize this emotional connection to unfamiliar people.[32]

Domestication and taming work hand in hand. The phylogenetic process of domestication renders individuals from domesticated species easier to tame. The details are still not fully understood but clearly include a change in the critical period for social imprinting.[33]

We notice this primarily when dogs seek relationships with our own kind, but dogs raised on a farm will readily seek out attachments with goats, sheep, or any other animal.

CONCLUSIONS

Exploring the general absence of dogs from zoos has enabled us to unpack just what dogs are and why we should not expect them to be on public exhibit. Domestic dogs are not absent from zoos because they are a product of artificial selection—they are in fact the outcome of natural selection. Dogs are not absent from zoos because they are supposed to live in human homes—most dogs do not live inside human residences. Finally, being from a domesticated (sub) species is not in itself what makes dogs tame, and thus domestication is not reason enough to keep dogs out of zoos.

Notwithstanding these confusions, we certainly do not advocate introducing dogs to zoos. Since most people are seldom far from dogs they can interact with as closely as they care to, there could hardly be much demand for exhibiting dogs. Instead, we believe these three common misunderstandings about dogs illustrate broader misconceptions about interpreting "the wild" and also prompt consideration of ways to provide dogs the best lives we can.

Although dogs may not be in zoos, most First World dogs are indeed captive animals. In some cases, such as dogs living in small apartments who may be taken out of the home only very briefly to complete toilet functions, they may be as confined as they would be if they were exhibited in zoos. First World dogs are not tolerated roaming free of a tether to a human. Most places, a leash of two meters (six feet) is required when a dog is off land owned by its keeper. Thus, even when outside the home, dogs are often unable to express any volition in their movements.

While there may be moves to change the label commonly used to refer to the human who controls a dog from "owner" to "guardian" or even "pet parent," it remains the case that dogs are legally property and may be transferred to another human at will and without consulting the animal's own preferences.

First World residents are often offended to see dogs roaming freely and "rescue" them from poorer countries where such dogs are common. This may occur even though dogs that have grown up living without human constraint will have considerable difficulty adapting to the life of a captive pet.[34] Recently, controversy arose in the United Kingdom when a planeload of street dogs and cats was airlifted out of Kabul, Afghanistan, in the chaotic final hours of the United States–led occupation of that nation—even as many thousands of desperate people were left behind to their fate at the hands of the Taliban.[35]

Do dogs living in free-ranging groups really need rescuing? Is a dog, neutered and deprived of almost all freedoms, trapped in a First World home, possibly alone for many hours a day, even in a cage barely larger than itself, actually living a more satisfying life than a free-roaming dog? A freely ranging dog may carry chronic diseases and have only half the life expectancy of a pet dog, but it is nonetheless free to roam and make its own life choices.

The widely accepted principles of animal welfare known as the "five freedoms" provide a useful lens through which to view the lives of pet dogs. Three of the freedoms could be termed "health" freedoms and are normatively well dealt with for modern First World pet dogs: freedom from hunger or thirst; freedom from discomfort; and freedom from pain, injury, and disease. The remaining two freedoms could be termed "psychological" and are much more questionable for pet dogs. These are the freedom from fear and distress, including mental suffering; and the freedom to express normal behavior, including access to sufficient space to perform behaviors as well as interact with conspecifics.[36] In our view, it is quite questionable whether First World pet dogs, even as they may be materially very well supported, are actually offered these psychological freedoms. Dogs are often left alone for extended periods, leading to psychological suffering, and are seldom given the opportunity to engage in natural behaviors such as hunting and scavenging, nor are they given opportunities for conspecific interaction including mating (figure 9.2).

FIGURE 9.2. A dog trapped in a home, as is common in First World countries.

Source: Photo by Наталья Демина. https://www.pexels.com/photo/a-dog-by-the-window-8601077/.

It is surely noteworthy that there is little movement toward the rescue of dogs from private homes. There is widespread acceptance of what, we submit, ought to be considered questionable and potentially harmful practices. One animal rights group, People for the Ethical Treatment of Animals, does argue that, "it would have been in the animals' best interests if the institution of 'pet keeping' . . . never existed,"[37] yet theirs is clearly a minority position. Even this group does not propose eradicating currently domesticated animals from homes.

This consideration of the forms of socially acceptable and unacceptable captivity to which dogs are commonly subjected naturally leads to a consideration of what a good life for a dog looks like. It must surely include veterinary care and healthful nutrition but should be broadened to include, among other things, freedom to express some control over their own lives. Rather than concerning ourselves with whether dogs can or should be truly wild, we suggest that our focus should be on aspects of dog welfare that are under our control.

Thinking about dog welfare and captivity in this way can in turn reorient discussion of the well-being of animals commonly kept in zoos. As Losos (this volume) cogently argues, evolutionary adaptation in animals in captivity can be remarkably rapid. Consequently, considerations of wildness may not be particularly helpful to maintaining a good life for animals in captivity. Rather, people charged with zoo animals' welfare should focus on the behavioral welfare of the animals in their care as evidenced in behavioral and physiological signs of stress and distress.

NOTES

1. See, e.g., "Best Friends with a Purpose: Cheetah Has a 'Therapy Dog' Companion at the Turtle Back Zoo," *CBS News*, 2020, https://newyork

.cbslocal.com/2020/02/03/cheetah-therapy-dog-turtle-back-zoo/; "A Cheetah at Cincinnati Zoo Has Become Best Friends with a Rescue Dog," *Newsweek*, 2019, https://www.newsweek.com/cheetah-cincinnati-zoo-best-friends-rescue-dog-companion-1470167; "Cheetah Cubs at the Columbus Zoo and Aquarium Get a New Playmate," *NBC News*, 2020, https://www.nbc4i.com/news/local-news/cheetah-cubs-at-the-columbus-zoo-and-aquarium-get-a-new-playmate/.

2. "Chinese Zoo Cheats by Disguising Hairy Dog as Lion," *Straits Times*, 2013, https://www.straitstimes.com/asia/chinese-zoo-cheats-by-disguising-hairy-dog-as-lion.
3. Wilfrid Blunt, *The Ark in the Park: The Zoo in the Nineteenth Century* (Hamilton: Tryon Gallery, 1976).
4. Nigel Rothfels, *Savages and Beasts: The Birth of the Modern Zoo* (Baltimore, MD: John Hopkins University Press, 2002).
5. "Zoo Owner Is Accused of Feeding Puppies to Snakes," *LA Times*, 1996, https://www.latimes.com/archives/la-xpm-1996-07-27-mn-28434-story.html.
6. "Zookeepers Fed Live Puppies to Starving Pythons in Front of Horrified Visitors," *International Business Times*, 2018, https://www.ibtimes.co.uk/zookeepers-feed-live-puppies-starving-pythons-front-horrified-visitors-1658403.
7. "Santa Barbara Zoo Introduces New 'Ambassador Dog,'" KEYT, 2019, https://keyt.com/news/2019/03/19/santa-barbara-zoo-introduces-new-ambassador-dog/.
8. Association of Zoos and Aquariums (AZA), *The Guide to Accreditation of Zoological Parks and Aquariums*, 2021, https://www.aza.org/accred-materials?locale=en.
9. For example, the San Diego Zoo (San Diego, CA) exhibits African painted dogs, the New guinea singing dog, and wolves; the Phoenix Zoo (Phoenix, AZ) exhibits the Mexican gray wolf, the coyote, and the maned wolf; the Brevard Zoo (Melbourne, FL) exhibits dingos.
10. Bradley P. Smith et al., "Taxonomic Status of the Australian Dingo: The Case for *Canis dingo Meyer*, 1793," *Zootaxa* 4564, no. 1 (2019): 173–97.
11. See Smith et al., "Taxonomic Status of the Australian Dingo"; and Clive D. L. Wynne, "Dogs' (*Canis lupus familiaris*) Behavioral Adaptations to a Human-Dominated Niche: A Review and Novel Hypothesis," in

Advances in the Study of Behavior, ed. Marc Naguib et al. (Academic Press, 2021), 52:97–162.

12. Charles Darwin, *On the Origin of Species by Means of Natural Selection* (London: John Murray, 1859).
13. Janet Browne, *Charles Darwin: The Power of Place* (Princeton, NJ: Princeton University Press, 2002).
14. Raymond Coppinger and Lorna Coppinger, *Dogs: A New Understanding of Canine Origin, Behavior, and Evolution* (Chicago: University of Chicago Press, 2002).
15. See Coppinger and Coppinger, *Dogs*; and Angela R. Perri et al., "Dog Domestication and the Dual Dispersal of People and Dogs Into the Americas," *PNAS* 118, no. 6 (2021): e2010083118.
16. Perri et al., "Dog Domestication."
17. See F. Hosseini-Zavarei et al., "Predation by Grey Wolf on Wild Ungulates and Livestock in Central Iran," *Journal of Zoology* 290 (2013): 127–34; Steven H. Fritts and David L. Mech, "Dynamics, Movements, and Feeding Ecology of a Newly Protected Wolf Population in Northwestern Minnesota," *Wildlife Monographs* 80 (1981): 3–79; and Alon Reichmann and David Saltz, "The Golan Wolves: The Dynamics, Behavioral Ecology, and Management of an Endangered Pest," *Israel Journal of Zoology* 51 (2005): 87–133.
18. Kathryn Lord et al., "Variation in Reproductive Traits of Members of the Genus *Canis* with Special Attention to the Domestic Dog (*Canis familiaris*)," *Behavioural Processes* 92 (2013): 131–42.
19. Kathryn Lord, "A Comparison of the Sensory Development of Wolves (*Canis lupus lupus*) and Dogs (*Canis lupus familiaris*)," *Ethology* 119 (2013): 110–20.
20. Howard S. Hoffman, *Amorous Turkeys and Addicted Ducklings: A Search for the Causes of Social Attachment* (Boston: Authors Cooperative, 1996).
21. Lord, "A Comparison of the Sensory Development of Wolves."
22. Oxford English Dictionary, "domesticate, v," https://www.oed.com/view/Entry/56668.
23. Arrian, *Arrian on coursing: The Cynegeticus of the younger Xenophon, translated from the Greek, with classical and practical annotations, and a brief sketch of the life and writings of the author. To which is added an appendix, containing some account of the Canes venatici of classical antiquity* (London: J. Bohn, 1831).

24. Simon J. M. Davis and François R. Valla, "Evidence for Domestication of the Dog 12,000 Years Ago in the Natufian of Israel," *Nature* 276 (1978): 608–10.
25. Jean Brixhe, *Le chien dans l'Egypte ancienne. Les origines* (Meretseger, 2019).
26. Andrew N. Rowan, "Global Dog Populations," *WellBeing News* 2, no. 5 (2020), https://www.wellbeingintlstudiesrepository.org/wbn/vol2/iss5/1.
27. Davis and Valla, "Evidence of the Domestication fo the Dog."
28. Rowan, "Global Dog Populations"; Raymond Coppinger and Lorna Coppinger, *What Is a Dog?* (Chicago: University of Chicago Press, 2016).
29. Coppinger and Coppinger, *What Is a Dog?*
30. Daniel G. Freedman, John A. King, and Orville Elliot, "Critical Period in the Social Development of Dogs," *Science* 133 (1961): 1016–17.
31. Erich Klinghammer and Patricia Ann Goodmann, "Socialization and Management of Wolves in Captivity," in *Man and Wolf: Advances, Issues, and Problems in Captive Wolf Research* (Dordrecht: Dr W. Junk, 1987), 31–59.
32. Christina Hansen Wheat et al., "Hand-Reared Wolves Show Similar, or Stronger, Attachment Toward Human Caregivers Compared to Hand-Reared Dogs," *bioRxiv* (2021), https://doi.org/10.1101/2020.02.17.952663; Nathaniel J. Hall et al., "Assessment of Attachment Behaviour to Human Caregivers in Wolf Pups (*Canis lupus lupus*)," *Behavioural Processes* 110 (2015): 15–21.
33. Alon Reichmann and David Saltz, "The Golan Wolves."
34. Sam Wollaston, "They Look Cute, but Should We Rescue Romania's Street Dogs?," *Guardian*, January 8, 2019, https://www.theguardian.com/global/2019/jan/08/they-look-cute-but-should-we-rescue-romanias-street-dogs; Erika Hobart, "Morocco Has 3 Million Stray Dogs. Meet the People Trying to Help Them," *National Geographic*, November 5, 2021.
35. Alannah Francis, "Afghanistan: Pen Farthing's Animals Saved from Kabul in 'Direct Trade-Off' with Refugees, Whistleblower Claims," *Inews.Co.Uk*, December 7, 2021, https://inews.co.uk/news/afghanistan-pen-farthing-animals-saved-kabul-afghan-refugees-whistleblower-1338925.

36. F. Ohl, F. van der Staay, and F. J. van der Staay, "Animal Welfare: At the Interface Between Science and Society," *Vetrinary Journal* 192 (2012): 13–19.
37. People for the Ethical Treatment of Animals (PETA), "Animal Rights Uncompromised: 'Pets,'" https://www.peta.org/about-peta/why-peta/pets/.

10

ZOO TIME

NIGEL ROTHFELS

> The electric light was turned on for the creatures in the Nocturama when real night fell and the zoo was closed to the public, so that as day dawned over their topsy-turvy miniature universe they could fall asleep with some degree of reassurance.
>
> —W. G. SEBALD

From the front of the huge Central Railway Station in Antwerp, a quick footpath leads to the entrance of the zoo. For more than a century, some seventy feet above the path, Josué Dupon's larger-than-life-size, 1899 bronze of a young man riding a camel, both now green with verdigris, has looked out over travelers, commuters, and daily streams of zoo visitors.[1] For the zoogoers, the proximity of the station is simply a convenience. Its nearness, though, points to shared histories. The great railway stations of the nineteenth century, along with museums of art and natural history, state opera houses, municipal waterworks and sewage systems, public libraries, city transportation systems, and much more, are all outcomes of the massive changes wrought in urban landscapes in the late nineteenth century as state revenues grew from industrialization, the rapid expansion of trade and transportation, and colonial plunder. Zoos and trains, in a way, sort of belong together—something quaintly echoed in the many small-gauge trains that still circle zoos, carrying passengers on trips of discovery in miniature carriages. Both the zoo and the train suggest distant places and even distant times. They mark dislocations—dislocations we all recognize as we exit unfamiliar stations, dislocations of animals shipped around the world for exhibition, dislocations modeled in bronze of a camel from the Asian steppes being driven by a youth from North Africa while both look out at a train station and zoo in Western Europe.

Perhaps the most prominent thing in any train station, anywhere in the world, is the station clock. In Antwerp it surveills the main hall, the so-called *salle des pas perdus*, regulating the steps of people, demanding, as train clocks and train watches everywhere have done since the end of the nineteenth century, that we all assent to "standard time." In many ways, how most of us think about time today is also an artifact of the nineteenth century, a consequence of cities, factories, trains, and schedules. Maybe it is not that surprising, then, that the lives of animals in zoos are also structured by nineteenth-century ideas of time, duration, and schedule—the temporalities of modern human life.

This is a simple observation, but I also think it tells us something important about the history of zoos, and maybe even their future. In this chapter I will explore how zoos have sought to manage the temporalities of animals so that they match our own. Using brief examples of nocturnal houses and feeding in zoos, I hope to suggest how a recognition of "zoo time" can lead to a deeper understanding of the challenges and potentials of "wilder zoos."

NIGHT TIME

In the 1950s, zoos began experimenting with nocturnal exhibits—and eventually whole nocturnal houses—in which exhibit lighting was reversed so that animals normally active during the night could be observed more easily by the public. In most cases, these exhibits focused on small mammals (figure 10.1). Between 1953 and 1965, nocturnal exhibits opened in Bristol, Chester, New York, Amsterdam, Cincinnati, Chicago, London, and Milwaukee. Antwerp Zoo opened its Nocturama in 1968. From these early institutions, and in the way these things usually progress, bigger, more exciting, and more complex nocturnal exhibits were developed over the following decades. In the United States, the most influential early trials were conducted in the basement of the old Lion House at the Bronx Zoo in New York in the spring of 1961. Using timers to turn lights on and off, the curator Joe Davis wanted to see if he could convince a pair of galagoes, small nocturnal primates from Africa, into thinking that day was night. The experiment worked, newspapers reported tongue-in-cheek on the newest "nightlife" spot in the Bronx, and by October of that year, a special room of the zoo's Small Mammal House was

FIGURE 10.1. Nocturnal Exhibit, the Small Mammal Building, Milwaukee County Zoo.

Source: Author photograph, 2021.

set up as the "Red Light Room." At 11 a.m., "night" started for the animals as the normal white lights in the room were turned off and red-filtered lights were turned on. Then, at 11 p.m., the white lights came back on.[2]

Throughout the latter half of the nineteenth century and first half of the twentieth, major zoos had regularly exhibited creatures like genets, fennec foxes, binturongs, civets, aardvarks, bats, spring-hares, porcupines, sugar gliders, and other creatures that would eventually inhabit the nocturnal houses. Frustratingly, the public saw little of these nocturnal animals because they simply hid in dark areas of their cages during the day. The Smithsonian's National Zoo in Washington did have an experimental nocturnal room in its Small Mammal Building in the late 1930s, but there, too, the exhibits were just box cages in a room, each kept covered to keep the light out.[3] Those who were interested—and one suspects they were mostly the curators themselves—could then lift the covers to see if they could see the animals moving about.

The reversed-light nocturnal exhibits that began to be developed in the 1950s and 1960s changed everything. For centuries, people had walked in front of exhibits of deer, antelopes, big cats, bears, and even elephants and giraffes, but watching a fishing bat swoop down and catch a fish, as the Bronx Zoo claimed one could see at their World of Darkness building, which opened in 1969, was something new. Still, from our perspective of more than fifty years later, the exhibits really don't seem like such a big deal, and, in fact, most are slowly disappearing. We lost interest because few of the exhibited animals truly grabbed the attention of the public. Curiosity about civets, for example, rarely extended beyond a story about expensive coffee. Calling fennec foxes "super cute" or making sure the audience had a good view of vampire bats drinking blood just wasn't enough after a few decades to warrant new buildings. Some of the most charismatic of the nocturnal animals, moreover, continued to be difficult to see, despite designing exhibits for low-light viewing (figure 10.2). The fishing bats may have been fishing and the flying squirrels may have been flying, but because the animals moved so quickly in their dim worlds, visitors complained that they couldn't really see the animals. Soon video screens were installed to slow down the movements of the animals.

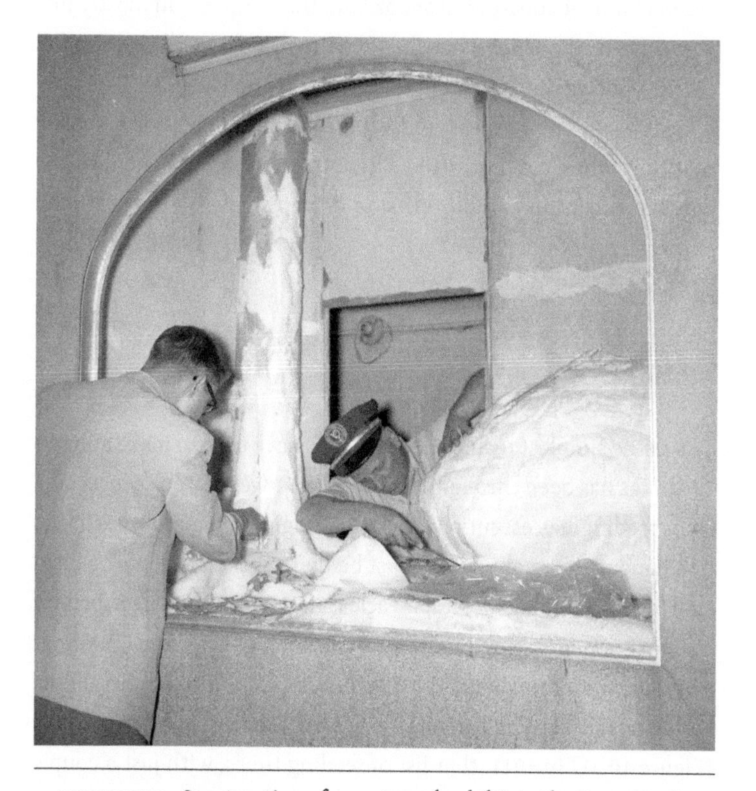

FIGURE 10.2. Construction of a nocturnal exhibit in the Bronx Zoo's Small Mammal Building, 1961. Most nocturnal exhibits were shallow in depth so animals could be more easily seen in the dim light. Still, visitors complained that they couldn't properly see the animals.

Source: © Wildlife Conservation Society, reproduced by courtesy of the Wildlife Conservation Society Archives.

Even if, in the end, the nocturnal houses were a sort of dead end for modern zoo designers, they point to how zoos have always sought to manipulate animals into becoming the right spectacle at the right time. Often, the manipulations are trivial. In cooler climates, for example, it isn't that uncommon for zoos to put large heated "rocks" in outdoor lion exhibits. The rocks are always placed in the perfect spot for the perfect photograph. Is that so bad? If the lion gets a nice warm chunk of concrete to snooze on, then I guess I'm mostly just happy for the lion. But what should we make of a gorilla exhibit full of trees, but the trees all seem excessively twined in vines? The public sees a verdant, natural landscape, but the gorillas learn quickly that the "vines" are hot wires. These trees are of no benefit to the gorillas; this is an exhibit decision driven solely by the aesthetic expectations of the public.[4]

FEEDING TIME

One of the simplest techniques that zoos have used to make animals do things has been through feeding. If lions were often seen as dull, their feeding has essentially always brought in the crowds. On the one hand, feeding obviously must be done at the zoo, but the decision to make feeding into a spectacle is not part of animal husbandry but part of generating public enthusiasm. Consider the "Feeding-Times of the Animals" in the 1911 edition of the *Illustrated Official Guide to the London Zoological Society's Gardens in Regent's Park* (Figure 10.3).[5] In 1911, that list of feeding times, with just a couple of additions over the years, had been in the guidebooks to the London Zoo for over fifty years. How did it come to happen that the pelicans at the London Zoo were fed at 2:30 p.m., seven days a week, for decade after decade after decade? Obviously, the birds' digestive systems did not evolve to be inundated with food once every day at

FEEDING-TIMES OF THE ANIMALS.

Pelicans	2.30 P.M.
Otters	3.0 P.M.
Polar Bears	3.0 P.M.
Eagles (except Wednesdays)	3.30 P.M.
Lions and Tigers	4.0 P.M.
Sea-lions	4.30 P.M.
Diving Birds in the Diving Birds' House	12 P.M. and 5.0 P.M.

N.B.—In November, December, January, and February, the Lions and Tigers are fed at 3 P.M., and the Sea-lions at 3.30 P.M.

FIGURE 10.3. Feeding schedule at the London Zoo, 1911.

Source: Zoological Society of London. Used by permission.

2:30 p.m. Having a reliable, official feeding schedule emerged out of the distinctive timed culture of the late nineteenth century. That the feeding schedule for pelicans, otters, polar bears, eagles, lions and tigers, sea lions, and penguins echoes a train schedule is clearly not coincidental.

Of course, even without a schedule, food has always been used to control animals. When the London Zoo opened in 1828, for example, one of the first exhibits was the bear pit—a hole in the ground with a substantial pole set into the middle. The bears would climb the pole, which had footholds along the way and was wrapped at the top with an iron strap on which was inscribed the motto "Excelsior," meaning higher. From the top of the pole—just above the head height of the viewing public—the bears would beg for food. The zoo operated a concession at the pit, and, when visitors wanted, they could buy buns to make the animals move. Two American black bears sent by the Hudson's Bay Company lived in the pit from 1829 into the 1850s, during which "they devoured an incalculable number of buns, provided by the liberality of visitors, for which they were always prepared to make an ascent to the summit of their pole."[6] The food was used to make the animals climb, make them beg, make them active at a specific time to meet human demands for when the animals should move. We are not so far from London's bear pit today. Anyone watching sea lions, dolphins, orcas, and many other animals being fed has experienced its modern analogs. Food remains a ubiquitous tool at the zoo—without it, everything from "free-flight" raptor shows, to up-close interactions with giraffes, to training an elephant to step into a squeeze chute for veterinary treatment would be a completely different kind of challenge.

Night time and feeding time are just two examples of the ways zoos manage interactions between animals, keepers, and the public between about 9 a.m. and 5 p.m. every day. The entire list of these techniques is long. For example, cathemeral animals (those that have no set time for most of their activities), along with crepuscular animals (those most active at dawn and dusk), vespertine animals (those most active in the evenings), and matutinal animals (those most active in the mornings), have almost always been restrained through tethering, stabling, or caging when zoos are closed. Despite elephants being naturally active both day and night, for example, historically they have been separated from one another and anchored by chains during the long night hours.[7] Similar sequestering has been used for other large animals like giraffes, dolphins, bears, and belugas and herd animals like antelopes. Most ungulates (everything from zebras to camels, rhinos, and manatees) and most carnivores (things like lions, tigers, hyenas, foxes, and wolves) are cathemeral, but in zoos, almost all are forced to abide by our diurnality—an activity pattern that is both relatively uncommon in the animal kingdom and becoming even more uncommon as animals seek to avoid humans. Behind the spectacular outdoor and indoor exhibits are the night cages and shift cages where the animals are isolated, including from one another, for more than half of every day.

Beyond night and day patterns, most other aspects of animal life at zoos are also closely managed. Births are organized with contraceptives, in vitro fertilization, sterilization, artificial insemination, sonograms, and caesarian sections. In the butterfly exhibits, chrysalids are lined up in order so that every day new imagoes emerge for the ever-expectant audience. Just what kind of system of production and consumption, what kind of regimes of biopower, make possible the constant churn of life in the butterfly exhibit? And, of course, every morning, before the public enters the serene butterfly exhibit, the dead from the day before are carefully collected and removed—most found tattered at the bottoms of windows. Of course, with "the beauty of nature" central to zoos more generally, any kind of "deformity," anything not conforming to curatorial and public expectations, is rarely tolerated. While senescence and death are sometimes acknowledged for the most well-known animals in the collection, the practice continues of replacing an animal with another before anyone notices. Between 1930 and 1960, for example, there were no fewer than five southern elephant seals named Roland at the Berlin Zoo.

A MORE NEEDED REVOLUTION

What do such examples have to contribute to our understanding of the form and future of a "wilder" zoo? Reflecting on this question, I'm struck by two queries posed by Ben Minteer and Harry Greene at the beginning of this volume. Ben asks whether we can imagine zoos where "wildness is not smothered but rather stoked" and points to two important historical "minirevolutions" toward wilder zoos: the Tierpark opened by Carl Hagenbeck near Hamburg in 1907, where animals were exhibited not in buildings filled with cages but in open-air, multispecies panoramas, and the immersive exhibits that began to be developed at the Woodland Park Zoo in Seattle in the 1970s, where designers sought to bring visitors into the exhibits themselves. Harry then asks whether zoos might be wilder and better if they at least try to "simulate participation in ecological and evolutionary processes."

In thinking through these questions, with their references to historical and future zoo revolutions, I'm struck, again, by the centrality of the human visitor's experience in everything about zoos. That isn't particularly surprising, of course: *zoos have always been built for people and not for animals.* This was as true at Berne's medieval bear pit as it was at Louis XIV's menagerie at Versailles and the Bronx Zoo's elephant house that opened in 1908. It remains true at the immersive Madagascar Rainforest exhibit at the Zurich Zoo, the just-because-we-can exhibit of whale sharks at the Georgia Aquarium, and even the *big* fantasies of the Danish Zootopia.[8] But saying that dodges the question raised by Jonathan Losos (a contributor to this book) in our discussions at the Arizona-Sonora Desert Museum: "Just what would a zoo built *for animals* actually look like?"

When Jonathan asked that, I responded glibly that the very question didn't make sense because zoos have only ever been built for people. As I have pondered the question, though, I have found myself reconsidering a form of revolution in zoos not mentioned by Ben, a quieter but more needed revolution that is associated with the mid-twentieth-century Swiss zoo director Heini Hediger and usually described as the field of "zoo biology"—a term suffering from the multiplied banality of its two words.[9] But however dull the name of the field, its promise is anything but trivial. Fundamentally, Hediger's work in zoo biology, building from theories of animal perspective developed by the German-Estonian Jakob von Uexküll, called on zoos to develop exhibits with a deep understanding of both animals' distinctive *Umwelten*—their unique perspectives or self-worlds—and their specific *Umgebungen*—the particular surroundings in which they live. In Hediger's approach, designers should consider how specific animals perceive and interact with the environments in which they actually live.[10]

Responding better to Jonathan's question, it isn't that zoos could or should be built for animals but that designers of a wilder zoo should always begin (and end) with questions about how animals might actually flourish in captivity. And returning to Harry's discussion at the outset, a wilder zoo would privilege the "messiness" of life and facilitate the full engagement of animals' uniquely formed ecological and evolutionary *Umwelten*, regardless of (or at least with less regard for) the convenience or comfort of the audience.

In a wilder kingdom, seeing nocturnal animals active at night might mean visiting the zoo during the night. A wilder zoo would not seek to fool either the animals or the public, would not use food coercively, or hide death, "imperfection," senescence, or disease. A wilder zoo would make it possible for the full lives of animals to be lived in the best circumstances a zoo could provide. A wilder zoo would not seek to control animal temporalities, to force animals into activity during "work hours," to speed up animals or slow them down, or essentially make animals "punch in" and "punch out" every day as they enter and leave their exhibits. Visibility should not be compulsory but chosen by the animals, and education at the zoo would stop being daycare for children and would help the public understand the full lives of animals.

In short, in a wilder kingdom, exhibits would be built around the needs of animals—including their needs around temporality. It is not that we should disregard the presence of humans at the zoo—zoos exist because of them, and they should be treated with respect. But that doesn't mean that zoos must be dumbed-down to the education level of children under ten, and it doesn't mean that zoos should be developed as just another kind of amusement park that considers the lives of the others-than-human as disposable, replaceable, and present only because they increase receipts, bring laughter, or even elicit "wonder."

NOTES

1. See Antoon Van Ruyssevelt and Fernand Schrevens, *Zoo Beeldig: Wandelen Langs Beelden en Gebouwen van ZOO Antwerpen* (Antwerp: MAS, 2007), 101.
2. Joseph A. Davis Jr., "Exhibition of Nocturnal Mammals by Red Light," *International Zoo Yearbook* 3, no. 1 (1962): 10. See also "Exhibiting Nocturnal Animals at the New York (Bronx) Zoo," *International Zoo Yearbook* 2, no. 1 (1961): 71. In a typical article titled "World of Darkness to Stir Nocturnal Animals to Life," in the July 31, 1965, edition of the *New York Times*, Edward Burks, for example, reported that "they are getting ready to build a 'far out' kind of night spot in the Bronx. This one will be swinging, jumping, wriggling, and creeping in the daytime, under conditions of artificial light."
3. The National Zoo experiment is described briefly in Sybil E. Hamlet, "History of the National Zoo," unpublished manuscript, 182–83, Smithsonian Institution Archives, Washington, DC.
4. See the exemplary historical work on primate enclosures by Violette Pouillard in "Structures of Captivity and Animal Agency, the London Zoo, ca. 1865 to the Present Times," in *Outside the Anthropological Machine: Crossing the Human-Animal Divide and Other Exit Strategies*, ed. Chiara Mengozzi (New York: Routledge, 2020), 40–57. My thanks to Dr. Pouillard, too, for being my guide to the history of the Antwerp Zoo.
5. P. Chalmers Mitchell, *Illustrated Official Guide to the London Zoological Society's Gardens in Regent's Park*, 9th ed. (London: Zoological Society, 1911), 110.
6. Philip Lutley Sclater, *Guide to the Gardens of the Zoological Society of London*, 21st ed. (London: Bradbury, Evans, 1868), 28–29.
7. See Brian J. Greco et al., "The Days and Nights of Zoo Elephants: Using Epidemiology to Better Understand Stereotypic Behavior of African Elephants (*Loxodonta africana*) and Asian Elephants (*Elephas maximus*) in North American Zoos," *PLOS One* 11, no. 7 (1916): e0144276.

8. See Ben A. Minteer, "The Parallax Zoo," in *The Ark and Beyond: The Evolution of Zoo and Aquarium Conservation*, ed. Ben A. Minteer, Jane Maienschein, and James P. Collins (Chicago: University of Chicago Press, 2018), 370–81.
9. For more on Hediger, who directed the Bern (1938–43), Basel (1944–53), and Zurich (1954–73) zoos, see Matthew Chrulew, "My Place, My Duty: Zoo Biology as Field Philosophy in the Work of Heini Hediger," *Parallax* 24, no. 4 (2018): 480–500.
10. See Riin Magnus, "Time-Plans of the Organisms: Jakob von Uexküll's Explorations Into the Temporal Constitution of Living Beings," *Sign Systems Studies* 39, no. 4 (2011): 37–56.

11

THE MICROBIAL ZOO

How Small Is Wild?

IRUS BRAVERMAN

> Every one of us is a zoo in our own right—a colony enclosed within a single body. A multi-species collective. An entire world.
>
> —ED YONG, *I CONTAIN MULTITUDES*

Does the Partulid snail care if she is at the London Zoo or at her place of origin in French Polynesia? Can a tardigrade distinguish between an urban backyard and the Amazon forest? And is there anything wilder, while also more human, than the microbiome within our guts, our skin, our eyes? The smaller the organisms are, the less wild they seem—and, for that matter, also the less unwild. In other words, the tinier the scale, the less relevant notions of wildness seem to be. Not many have bothered asking, for example, whether bacteria on our nail or an ant on our desk is wild or not. Size is an important yet much neglected aspect of how we imagine the wild, as well as how we imagine agency.

This, arguably, is why zoos—which usually exhibit what they conceive as wild animals—have not typically been in the business of exhibiting microorganisms. I would like to invite us to envision just how wild it would be for a zoo to focus precisely on such microscopic dimensions of life. This was the zoo of my dreams, until I discovered that it already existed in the world (now how wild is that?!). It is called Micropia-ARTIS (Micropia for short) and is situated in Amsterdam. To the best of my knowledge, this is the only institution of its kind in the world. Micropia presents an opportunity to think more critically about wildness and the governance of more-than-humans.

This chapter draws on in-person interviews I conducted with Micropia's two microbiologists (one of whom is also the chair of this institution) and on onsite observations in the summer of 2019 to illuminate some of the broader questions about how zoos might envision a wilder future. I will start by considering the status of Micropia: is it a museum, a zoo, a lab, a research facility—or a hybrid of these institutions? Next, I will contemplate the project of making microbes live—as well as making (and letting) them die. Here I draw on Michel Foucault's terminology, and his work on biopolitics in particular,[1] to highlight the value systems and priorities that are engrained in and at the same time invisibilized through this project. Choice, isolation, visibilization, and attraction are all part of the "theater of production" that takes place at Micropia, rendering microbes governable, lovable, and also killable.

MICROBES: AN INTRODUCTION

Until recently, our planet was thought to be inhabited by nearly 10 million species. Based almost solely on species that can be seen with the naked eye, this estimate did not account for smaller species such as bacteria, archaea, protists, and fungi. Yet these microbial taxa are the most abundant, widespread, and longest-evolving forms of life on the planet.[2] Recent studies suggest, along these lines, that Earth might be home to a staggering 1 trillion species.[3] For example, one single gram of agricultural soil can routinely contain more than ten thousand species. Similarly, the human microbiome is made up of nearly 10 trillion bacterial cells.[4]

A recent surge in research has also revealed that microbiota residing on and inside organisms profoundly influence animal health. Some have thus called for a "microbial renaissance of conservation biology," where "biodiversity of host-associated microbiota is recognized as an essential component of wildlife management practices."[5] Relatedly, conservationists are realizing that when a host becomes extinct, "all the life forms that are associated with it which are definitely unique—they are gone as well."[6] Beyond the bodies of animals (and plants), microbes are also part of Earth's crust, atmosphere, ocean, and ice caps. In total, the estimated number of microbial cells on Earth is estimated at nonillion, which exceeds the estimated number of stars in the universe.[7] It is therefore somewhat surprising that, with few exceptions,[8] so little attention has been devoted to microbes in museums and zoos.

Micropia is a rare, if not a single, exception. Identifying itself as "the only museum of microbes,"[9] Micropia seeks to educate the public on the importance of microbes for the interconnectivity in the natural world. Micropia's "Intention and Purpose" statement describes a "serious knowledge gap between the science and the general public" in this context, which is dangerous because the "unknown is unloved."[10] By contrast, Micropia depicts microbes as the "most successful organisms," given their critical function for all forms of life. The goal of the institution, to bring "nature startlingly close," highlights Micropia's foundational belief that microbes are, indeed, part of nature. Moreover, they are considered to be nature's foundation. In the words of Micropia's chair: "[It] all started with microbes. . . . If we didn't have microbes, nature wouldn't exist, we wouldn't exist."[11] The question that remains to be discussed is whether, in addition to being natural, microbes are also considered wild.

Micropia exhibits more than 240 different microbe species,[12] aiming to let the general public "see them with their very own eyes." Many of Micropia's displays are accessed through a 3D viewer attached to a microscope via cameras on the eyepieces of the microscope. According to Micropia's website, this setup is intended to give the visitor a feeling of "diving into the invisible world,"[13] and in this sense it is not unlike immersion exhibits in zoos.[14] More broadly, and despite presenting itself as a *museum* of microbes, Micropia shares certain features with zoos. The most obvious common theme is that both are collections of living things. The microbes are contained within glass slides and in this respect are similar to zoo animals who are contained within cages and enclosures. Moreover, both types of exhibits seek to promote a love for and connection to the natural world, spurning an interest in "protect[ing]" and "enrich[ing]" it.[15]

Nonetheless, Micropia's chair, Remco Kort, and the institution's microbiologist Jasper Buikx were adamant that Micropia should not be considered a zoo. Buikx explained that "having a life collection in a museum setting is something completely different from having it at a zoo, because [we] must *keep* everything alive. So, that's why we also have our own lab that continuously maintains the collection."[16] Kort tended to agree with Buikx that Micropia is more a museum than a zoo because of the active, "life-support" nature of the effort and the overwhelming reliance on technology, but he also agreed with my proposal to see Micropia as a hybrid institution that travels between zoo, museum, science center, and lab. This of course raises interesting questions about the nature of zoos: perhaps zoos simply mask from their visitors the degree to which they, too, "keep everything alive"? The aquarium is another institution that complicates the distinctions, as its watery environment usually requires intense maintenance.[17]

MAKING MICROBES LIVE: A THEATER PRODUCTION

To exhibit microbes, Micropia must undertake two initial tasks: first, to keep the microbes alive and, second, to make them visible to human visitors. A third task that takes place later in the process is to make the microbes attractive to visitors. As Kort explained: "from a morphological point of view, bacteria are not the most exciting animals. Most of them just look like spheres or beads." Nonetheless, he feels a strong emotional bond with these organisms. He described how, after a while, one can distinguish different behaviors in different bacteria. In his words: "if you work with a microorganism for a very long time . . . you feel strongly connected to it." This, again, is in line with how zookeepers relate with their animals and possibly also how some scientists in research labs operate.[18]

When applied to microbes, these managerial tasks can only be carried out through technology—microscope and DNA maps in particular. The Dutch scientist Antonie van Leeuwenhoek is credited with discovering and describing what are now defined as bacteria and protozoa. He was a microscope maker and first found bacteria in rainwater and in the plaque between his teeth. He called the tiny organisms "animalcules," betraying an assumption that they must fit into the human, animal, or plant categories. As microscope technology progressed, so did the scientific understanding of microbiology. In the 1800s, biologists learned about cells, realizing that the animalcules that van Leeuwenhoek observed were single-celled organisms, unlike larger multicellular organisms.

What does keeping such unicelled organisms alive entail in the context of an exhibition? First, it entails a choice. Kort explained that pathogens, GMOs, and extremophiles are not contenders from the outset given the biosecurity, regulatory, and physical challenges that keeping them alive would entail. Next, the microbe must undergo a process of isolation from all other organisms. "That's lab work . . . you need to be able to cultivate them," Kort told me. "The life form of the cultured cell is a manifestly technological one," the STS scholar Hannah Landecker explains along these lines. She continues:

> It is bounded by the vessels of laboratory science, fed by the substances in the medium in which is bathed, and manipulated internally and externally in countless ways from its genetic constitution to its morphological shape. Its existence bears little resemblance to the body plan or the life span of the organism from which its ancestors were derived. Contemporary life in this particular form is something that exists and persists in the laboratory, the niche of science and technology.[19]

In addition to the isolation of strains and their culture on Petri dishes decontextualized from their organic life, exhibiting microbes also entails what Kort likens to a "theater production" (figure 11.1).[20] In his words, this is "the stuff that we need to be able to *show* them—to keep them alive on display for a certain number of hours." Such a visible display requires "a very interesting science that is specific for our microbe museum." As a scientist, he explained, you are not too concerned about that. But Micropia cannot replace its exhibitions every ten minutes, he continued. Producing a meaningful visual exhibit is especially difficult because of the microbes' relatively short life span. According to Kort: "Bacteria have a generation span of about six minutes. So in ten hours, you will have a billion bacteria from one single bacteria. Not only do they have a short generation span, but they also take up pieces of DNA from the environment and exchange it with each other. So they are continuously changing their DNA. . . . That's [why] bacterium evolves a million or even a billion times faster than humans."

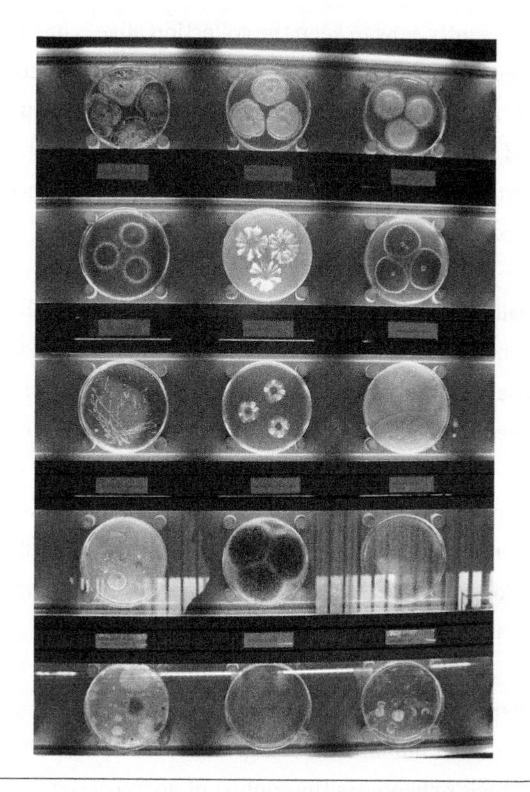

FIGURE 11.1. Micropia's petri dish exhibit of bacteria.

Source: Photo by the author, June 2019.

Additionally, "many of them don't like light, many don't like to be in this enclosure where the oxygen is deprived. So, that's a whole science by itself. This is why our lab mimics those conditions, and we are constantly working on how we can improve that, to see how we can add nutrients to see that they are doing better."[21] The existence of the lab alongside the exhibit serves to provide a continuous supply of microbes, bearing interesting similarities to captive-breeding facilities in zoos.

Put differently, how to keep the collection alive is one thing, but to keep the collection alive *on display* is a challenge at a completely different magnitude and is unique to Micropia. Jasper Buikx explained that doing this requires technical and technological adjustments. For example, the design of new types of cover slides. In his words:

> Normally, you slide one or two drops of the sample and then you have another piece of glass on top. This works fine in the lab because you often just need a couple of minutes to do whatever you need to do and then you dispose of the slide. But here, of course, we don't have time to continuously change the slide in our thirty different exhibits. So we designed a new cover slide that contains much more fluid. That way, it stays healthy for much longer, which [saves] us from changing everything every ten minutes.[22]

The element of visibility is important in Micropia and resonates with the centrality of humans' "looking" in zoo exhibits,[23] discussed in Nigel Rothfels's chapter in the context of nocturnal design.

In addition to showing visitors the microbes, they must also be made attractive. One of Micropia's more attractive exhibits was the "kiss-o-meter." Kort explained the evolution of this exhibit: "We did that research initially. I carried it out here in the zoo about bacterial transfer through French kissing. So, we had couples give French kisses and I was standing there with my stopwatch and they had ten seconds and then we sampled the saliva. We did a tongue swap and characterized the microbiota. We used it in Micropia. . . . This goes to give you some idea about how many microbes are transferred."

In ending this contemplation about Micropia's exhibition of life, I would like to circle back to isolation as a shared theme between Micropia and zoos. Like many modern zoo collections, the bacterial strains are captive bred and are not obtained from "natural" (wild?) strains. "In nature you rarely see pure cultures," Kort explained. Under those conditions, bacteria form communities, and, within those communities, different cells take specific properties. In Kort's words: "one becomes a swimmer and leaves the community, another becomes a spore former, and the third transforms into a different function." In this sense, bacterial communities act like a multicellular organism. This behavior was not observed in laboratory isolates, however. There, "bacteria are under these artificial cultivation conditions for so long that they lose this complex behavior."[24]

While such discussions of purity, isolation, and the effects of domestication occur in zoos too, Micropia is different in that it does not regard the aftermath of the exhibit as a central concern. In zoos, what to do with animals who can no longer be exhibited for various reasons—and how to euthanize them, in particular—has been a topic of longstanding ethical debates.[25] As I discuss next, in Micropia this is much less the case.

MAKING MICROBES DIE

Like in zoos, the art of making (organisms) live—in the Foucauldian sense of managing their health and populations—has a less pretty side: making die. Yet unlike in zoos, where the ethics of killing/culling and euthanasia has become central to animal management, in Micropia this is not a central consideration. As a starting point, my interviewees clarified that despite their care toward these organisms, they do not see them as animals, and in some cases, such as viruses, they don't even consider them to be living beings. This goes back to the understanding that what humans cannot see with their naked eyes they tend to not recognize as ethical subjects. In the process, such organisms are made "killable."[26]

What are Micropia's practices and ethics of killing microbes? When I asked him about this, Kort dismissed the question by explaining that each person constantly "kills microbes by swallowing them." Once they enter your stomach, he told me, "you kill all, or at least most, of them. Trying to avoid that kind of thing is very hard." By being human and living your life, you are already killing bacteria all the time, he explained more broadly. Buikx added that "all the lab work ends up with putting the microbes into an autoclave." An autoclave, he explained, "is a machine—like a pressure cooker—that allows sterilization. A regime of twenty minutes in 20 degrees and then nothing is alive anymore." As he listened to this description by his colleague, Kort asked to clarify that "the primary focus is always to keep microbes alive and have them multiply." This, at least, is the case at the museum, the two clarified. However, at the lab that supplies the museum with the specimens, killing is the routine and making live the exception.

Nonetheless, even at the museum level, "I guess at some point the stuff you introduce, you also have to expose," Kort tells me, adding: "the amount of stuff that we put in the autoclave is just 1 or 2 percent—it is very minute. The rest either dies by itself, or you are trying to keep it alive for as long as possible by adding nutrients [and] creating the perfect surroundings using glow lights and stuff. Since bacteria are cultivated under artificial conditions—under a cover slip and a microscope—that's a form of containment." What is implied is that once the special conditions are not provided, the microbes die on their own.

CONCLUSION

Micropia is an illuminating case study about what it means to create and maintain a living collection. It highlights aspects that might be less obvious in the zoo context, such as the need to isolate collections and the "theater" production around their exhibit. At the same time, it also highlights the uniqueness of bacteria and the biopolitical assertions about what is worth exhibiting—namely, "making live" through making visible—as well as what happens behind the scenes when visibility is no longer required. The microbes are enlisted by Micropia to labor for a purpose: exposing the less visible microaspects of natural life. Yet, the ethics are more akin to labs rather than zoos, as this is where most of the work on microbes has taken place until now.

I would like to end by underlining two themes: first, the differences between natural and wild. While microbes are certainly considered "natural," and, as such, they are exhibited as worthy of our attention and love, they are not really thought of as "wild" per se. I would like to pause to consider this distinction and its implications for both how we think about zoos and also how we think about microbial life. It is the size of their life's form as well as their morphology as single-cell communities that seem to distinguish bacteria from forms of life that are seen as worthy of more comprehensive ethical considerations.

The second theme underlined here is the collapse of institutional distinctions. Micropia illuminates that once we see life as a complex ecosystem rather than in its individual manifestations, then life's multitude unravels, collapsing the institutional boundaries that have so carefully been constructed over the decades between the natural history museum, the zoo, the lab, and the research facility. This project's hosting venue, the Desert Museum, sets out to collapse the zoo–natural history museum–botanical garden distinction in significant ways.

Yet we still have a long way to go before fully realizing that wildness manifests in small things no less than in big ones. To paraphrase Arundhati Roy's *The God of Small Things*: we may want to look inside ourselves to discover just how wild we already are. The zoos of our transformative time would be an excellent place to start.

NOTES

1. Michel Foucault, *The Birth of Biopolitics*, ed. Michel Senellart (New York: Palgrave Macmillan, 2008).
2. Jay T. Lennon and Kenneth J. Locey, "There Are More Microbial Species on Earth Than Stars in the Galaxy," *Aeon*, September 10, 2018, https://aeon.co/ideas/there-are-more-microbial-species-on-earth-than-stars-in-the-sky.
3. Kenneth J. Locey and Jay T. Lennon, "Scaling Laws Predict Global Microbial Diversity," *PNAS* 113, no. 21 (2016): 5970–75.
4. Ron Sender et al., "Revised Estimates for the Number of Human and Bacteria Cells in the Body," *PLOS Biology* 14, no. 8 (2016): e1002533.
5. Brian K. Trevelline et al., "Conservation Biology Needs a Microbial Renaissance: A Call for the Consideration of Host-Associated Microbiota in Wildlife Management Practices," *Proceedings of the Royal Society B* 286, no. 1895 (2019): abstract.
6. Remco Kort, professor of microbiology at the Vrije Universiteit Amsterdam and holder of the ARTIS-Micropia Chair, ART-Micropia, interview by author, June 24, 2019, Micropia, Amsterdam, the Netherlands.
7. See, e.g., Ed Yong, *I Contain Multitudes* (New York: Ecco, 2016), 1, 235. Here, the microbes are not exhibited but are studied as part of creating a better exhibit for marine mammals, dolphins in particular.
8. See, e.g., the Shedd Aquarium and San Diego Zoo, as discussed in Yong, *I Contain Multitudes*; and in my interviews with Bill Van Bonn, director of animal health at the Shedd Aquarium.
9. "ARTIS-Micropia," Micropia, https://www.micropia.nl/en/.
10. See, e.g., "Love the Unloved," by Jasper Buikx, https://www.youtube.com/watch?v=Uxq2NAPQlp8.
11. Kort, interview.
12. From Kort: "We want to call something a species, strain, or whatever. [But] that is something so fluid in the microbial world. Something we

might call species is so diverse and changes so quickly that it's . . . irrelevant in the bacterial world. It is so fluid, it's so different that even within species we have different traits." Interview.

13. "How Micropia Came About," Micropia, https://www.micropia.nl/en/footer/about-micropia/how-micropia-came-about/.
14. On the nature of immersion exhibits, see Irus Braverman, *Zooland* (Stanford, CA: Stanford University Press, 2012).
15. "Intention and Purpose," Micropia, https://www.micropia.nl/en/footer/about-micropia/intention-purpose/.
16. Jasper Buikx, microbiologist at ARTIS-Micropia, interview by author, June 24, 2019, Micropia, Amsterdam, the Netherlands.
17. On the relationship between zoos and aquariums in terms of conservation management, see Irus Braverman, "Fish Encounters: Aquariums and Their Veterinarians on a Rapidly Changing Planet," *Humanimalia* 11, no. 1 (2019): 1–29.
18. Irus Braverman, "Gene Drives, Nature, and Governance: An Ethnographic Perspective," in *Gene Editing, Law, and the Environment: Life Beyond the Human*, ed. Irus Braverman (New York: Routledge, 2017), 54–73.
19. Hannah Landecker, *Culturing Life: How Cells Became Technologies* (Cambridge, MA: Harvard University Press, 2007), 3. See also "History of the Cell," *National Geographic*, 2019, https://www.nationalgeographic.org/article/history-cell-discovering-cell/8th-grade/; "Animalcule," *ScienceDirect*, 2021, https://www.sciencedirect.com/topics/agricultural-and-biological-sciences/animalcule; "History," BBC, 2014, http://www.bbc.co.uk/history/historic_figures/van_leeuwenhoek_antonie.shtml; Carrie Arnold, "The Man Who Rewrote the Tree of Life," *Nova*, April 30, 2014, https://www.pbs.org/wgbh/nova/article/carl-woese/; "Are Bacteria Animals or Plants?" Australian Museum, 2021, https://australian.museum/learn/species-identification/ask-an-expert/are-bacteria-plants-or-animals/.
20. Kort, interview.
21. Kort, interview.
22. Buikx, interview.
23. See, e.g., John Berger, *About Looking* (New York: Vintage, 1980), 3–30; and Irus Braverman, "Looking at Zoos," *Cultural Studies* 25, no. 6 (2011): 809–42.

24. Kort, interview.
25. Irus Braverman, *Zoo Veterinarians: Governing Care on a Diseased Planet* (New York: Routledge, 2021), chap. 4.
26. Judith Butler, *The Force of Nonviolence: An Ethico-Political Bind* (London: Verso, 2020).

12

A HOME FOR THE WILD

Architecture in the Zoo

NATASCHA MEUSER

The construction tasks specific to zoological gardens and aquariums have long since developed into a distinct branch of architecture carried out by specialists and expert planners. What distinguishes these professionals is their ability to think in an interdisciplinary manner and their recognition of the importance of zoological expertise. Recent examples show that facilities are particularly successful when architects consider the concerns of zoologists, visitors, and animals alike.

At the same time, society's concept of the optimal coexistence of humans and animals has undergone a fundamental change. The realization that animals are not mere objects for display but beings with rights of their own is becoming more established. This change in the assessment of an optimal or at least appropriate coexistence of humans and animals can be read in the historical development of the architecture of the zoo. As Ben Minteer (this volume) writes, the zoological garden has changed from a living trophy collection, to a museum with living exhibits, to a naturalistic adventure park with a moral mission.

In this context, zoo architecture can be understood as the visible examination of the relationship between humans and animals. How to support the mission of species preservation in zoos and a morally justifiable human-animal relationship through building culture is a task by which future generations of zoological gardens will have to be measured—whether it is the numerous novel foundations of new parks, not least of the Asian region, or the sometimes bitterly necessary modernization of existing buildings, in which monument-preservation aspects increasingly must be taken into account.

Architecture is given an additional educational task in that it is supposed to help show in the zoo which ecological connections lead to the functioning of the world as a whole. Beyond a fun experience space, modern zoo architecture, like a modern museum, must encourage reflection and, above all, action: How valuable is nature (including wild nature) to me—and what can I do to conserve it myself? In this way, the zoo acts as an interface between the experience of visiting the zoo, the living animal, and science and its communication. A "wilder zoo," in other words, is inescapably a design discussion, one that must acknowledge a wider ecological and societal context.

AQUARIUMS AS A NEW STAGE FOR CLIMATE PROTECTION

Since the United Nations Framework Convention on Climate Change was adopted in Rio de Janeiro in 1992, the scientific recognition of and demand for intact ecological cycles has grown into a global movement. All over the world, the consequences of climate change have become tangible. As permafrost areas in Siberia begin to thaw in summer and forest fires in the Mediterranean become an annual tragedy, scientists are sounding the alarm ever louder about melting polar ice caps and the resulting rise in sea levels. Some researchers now predict that melting freshwater from Greenland's glaciers could lower the salinity of the Atlantic Ocean. According to hydrologists and climate scientists, if this scenario were to occur, the Gulf Stream could come to a standstill, significantly altering the weather in Europe.[1]

We are ignorant about the oceans, which European culture has always viewed only from the margins. When the conquerors and colonialists conquered the open seas from the fifteenth century onward, they encountered an inhuman world. On the water, the adventurers were at the mercy of nature. Men had more respect for the sea than for the peoples they encountered on the new coasts. It is thus no surprise to note that we have only explored a fraction of the water that surrounds us. It was not until the mid-twentieth century that the first heroes in diving bells ventured into the depths of the oceans—not with submarines hunting for targets but with an engineer's motivation to test the limits of what was technically possible. What deep-sea divers like Jacques-Yves Cousteau and other pioneers found were completely unknown, wild worlds that presented themselves in impressive images as colorful as a meadow of flowers. The sea seemed to become more conquerable with every ever more daring exploration.

Cousteau, who shaped our image of the sea with his more than one hundred movies about underwater worlds, was also one of the first researchers to speak out for its preservation. His Cousteau Society was committed not only to the exploration of the oceans but also to their protection. That was in 1973, inspired by the Club of Rome and its report on the limits to growth. Since then, marine research has always been linked to concerns about increasing pollution. Especially since the use of satellites has also become possible for civil institutions and nongovernmental organizations, almost every corner of Earth has been under constant observation, including the oceans.

What the photos and measurements tell us about the changes in nature caused by human intervention is both revealing and sad. Besides the rising water temperatures, which lead to the extinction of species because of the resulting decrease in oxygen content, other environmental problems have become visible: overfertilization of the oceans through nitrogen input via groundwater, streams, and rivers during intensive farming; plastic waste that is washed unfiltered into the seas; and fishing nets that are carelessly thrown overboard on the high seas and become deadly traps for marine life. The list of environmental sins caused by humans could go on.

Given the scientific admonitions that are repeatedly formulated around the globe, and also given the clearly visible dependence of the climate on the cleanliness and regenerative capacity of the world's oceans, the topic of water has advanced to become the central issue when it comes to the future viability of our planet. The new interest in the oceans and seas is thanks to the increased environmental awareness and the realization that the emission of greenhouse gases also directly damages the world's oceans.

This is reflected in the information available on the internet and on television—but also in the popularity of maritime educational institutions. Marine museums and aquariums are experiencing a new boom not only in terms of visitor numbers but also in terms of the quantity of buildings. They are increasingly taking on the task of making their visitors aware of the goals of climate protection policy and improving overall public knowledge. It's an example of how a concern about "the wild," in this case, the health and sustainability of the planet's marine systems, has become part of the experiential and educational environment of modern zoological parks.

Since 1990, the number of public display aquariums has tripled. In China alone, the number of aquarium houses has increased from six in 1991 to 225 today, reflecting their popularity with visitors. According to estimates by the World Association of Zoos and Aquariums (WAZA) and members of the International Aquarium Network (IAN/IAC), some 700 million people visited an aquarium in 2018, including 150 to 200 million in China. This number is not surpassed by any other group of conservation-oriented institutions.

New houses are continuously being built, and outdated exhibition concepts are being adapted to new educational goals (figure 12.1). It is noticeable that aquariums are rarely advertised with their fish population but mostly with structural superlatives: the

FIGURE 12.1. Conceptual collage for the Research Center of the Seas of Cortez in Mazatlán, Sinaloa, Mexico, by Tatiana Bilbao ESTUDIO. Completion date: 2022. Designed by the Mexican architect and university lecturer Tatiana Bilbao, this building is perhaps the most recent example of the trend toward the wildness of zoo architecture seen in the Americas. The Marine Research Center is a place where the Pacific coastal landscape meets human civilization. It seems as if nature has assimilated the building. The aquarium is part of a regeneration plan for the Sea of Cortez, described by the famous oceanographer Jacques Cousteau as "the aquarium of the world." This natural area is one of the most biologically diverse bodies of water on Earth.

Source: Courtesy of Tatiana Bilbao ESTUDIO.

largest capacity of a tank, the thickest pane of glass in a tank, or the longest walkthrough glass tunnel. Measured by the frequency of its users, however, construction tasks are still underrepresented in the debate on contemporary architecture, and literature on zoo and aquarium buildings is rare. This makes it all the more important to highlight their peculiar character in order to provide the designing architect and his client—the zoologist—with planning parameters and quality criteria for a sustainable building.

COLUMN-FREE HALLS AND JOINTLESS PANES

The focus in zoos and aquariums is on paying visitors. It is a concern that has understandably been exacerbated by the global pandemic and its devastating impact on visitation, which has put many zoos in dire economic straits. To cite just two examples, the Munich Zoo was closed for 116 days in Corona Year 2020. Instead of 2.72 million people as in 2019, only around 750,000 came. The impact on visitor numbers at the zoo and aquarium in Berlin is similarly drastic. Here, a drop in revenue of around 12.66 million euros was recorded in 2020. For 2021, revenue losses of the same amount were expected.

In addition to these challenges, the zoo continues to face an institutional dilemma. On the one hand, it needs to live up to its responsibility and create an awareness of nature, and on the other, it also seeks to create a pleasant setting for visitors that offers childcare, travel substitutes, and event gastronomy at the same time. In this context, the focus primarily on children slows down the zoos' ability to transform. In other words, as long as the zoo lures visitors with playground equipment and attractions and only wants to indirectly reference nature conservation, it will take at least twenty years until these children can implement a new environmental awareness and responsibility in their personal and professional adult lives.

What does this mean for efforts to rethink the zoo—as we are doing in this book—as a more ecological and transformative space (even if not really "wild")? For starters, it means a greater emphasis on consciousness-raising about animals and environmental protection and less time on amusement. Theo Pagel, director of the Cologne Zoo, has already said as much. "For us, the motto is 'walk the talk': we have to back up our words with actions. That means zoos should be role models in terms of ecology, right down to their gastronomic offerings. You can't preach environmental awareness and then sell cheaply produced stuffed animals in the zoo store."[2]

Zoos increasingly recognize this. That is, they don't want to be just theme parks. In fact, a noticeable process of transformation is currently taking place in large zoos in German-speaking countries. With regard to their structural infrastructure, three points can be summarized:

1. Zoos are trying to identify the best possible design via architectural competitions. The debates are sometimes passionate, and not only among experts. The building task itself benefits most from this, as it also leads to architects beyond the established offices becoming interested in and qualified for this highly specific design task.
2. When it comes to the question of further increasing the zoos' range of uses by adding facilities that attract the public or a playground, the focus today is on animal welfare, not visitor welfare. While the zoo used to be a place for Sunday concerts or adventure playgrounds, today the seriousness of species preservation has moved back to the forefront. This can also be deduced from the numerous architectural competitions, in which the latest findings

of species-appropriate husbandry and modern museum education are considered.

3. Architecture itself has gained new importance in zoo marketing. Spectacular indoor buildings such as Gondwanaland in Leipzig or the elephant enclosure in Zurich have led to a new architectural self-image of zoos. In addition, zoo directors have recognized the preservation of valuable building fabric as a location factor alongside the preservation of the animal population.

In order to integrate the new buildings into the existing zoo landscape, zoological gardens have begun to draw up frameworks and strategic plans (see figures 12.2–12.4) in which the mostly public

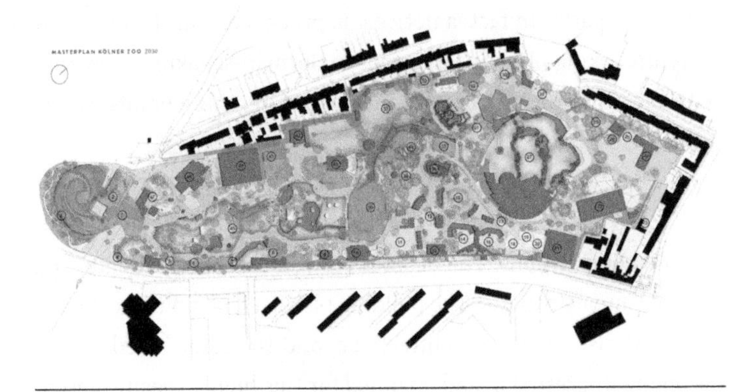

FIGURE 12.2. Cologne Zoo Master Plan, Cologne, Germany. Masterplan 2030. Planning: Zooquarium 2020. The inner-city location does not allow the Cologne Zoo to build large-scale new buildings. Against this background, the handling of existing buildings comes to the fore. As recently as October 2021, the Südamerikahaus, one of the historic buildings dating from 1899, was renovated and converted. The Cologne Zoo underlines its educational claim with, among other things, the construction of a so-called Green Classroom in the middle of the zoo grounds, as well as the conversion of the former director's villa into the multifunctional Villa Bodinus for event gastronomy and functions.

Source: Courtesy of Cologne Zoo.

investments of up to more than 100 million euros are presented transparently to the municipal parliaments. While Magdeburg's Master Plan 2025 only provides for seven million euros in investments over the next four years, the cost forecast for the Cologne Zoo (Master Plan 2030, 127 million euros) corresponds to the construction of an entire residential quarter. As Pagel puts it: "The zoo still has its justification. Due to advancing urbanization and digitalization, people have a great need to experience nature and animals

FIGURE 12.3. Magdeburg Zoo master plan. Magdeburg, Germany. Framework plan 2025. Planning: Meuser Architekten GmbH, 2021. The zoo management has had an outline plan drawn up that visualizes and coordinates the planned investments for the coming years. The focus is on optimizing animal husbandry according to current scientific findings. Four years of intensive planning are scheduled up to 2025 to renovate, convert, and expand individual areas. The aim is to strengthen the core competencies of species conservation, recreation, education, and research; to further improve working conditions, safety, and animal welfare; and thus to secure the future viability of the Magdeburg site as an internationally important zoological institution. The measures will be accompanied by renovations to the technical infrastructure and will take place during ongoing operations. A particular focus is on the renovation of the tiger and chimpanzee enclosures. The large enclosure for African elephants will be extended to accommodate young bulls.

Source: Courtesy of the author.

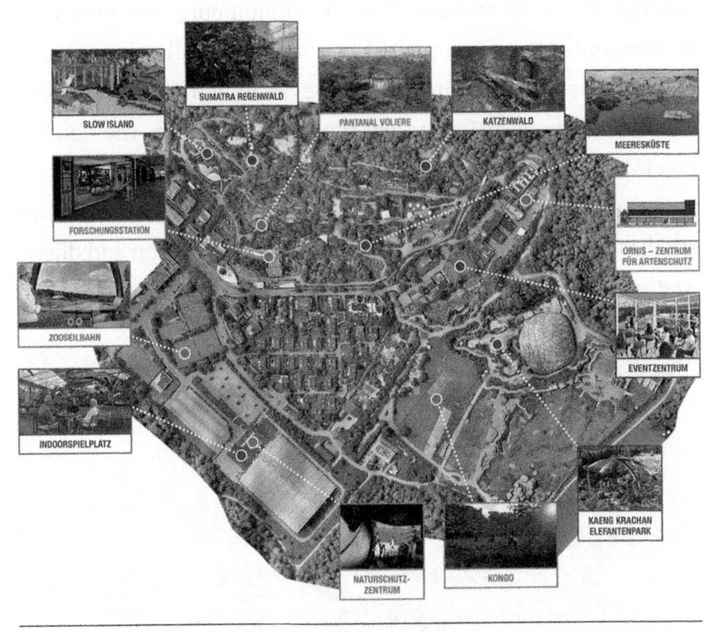

FIGURE 12.4. Zoo Zurich development plan. Zurich, Switzerland. Development plan 2050. Planning: Zoo Zurich/Severin Dressen 2021. Since the legendary zoo director Heini Hediger (1908–1992), Zoo Zurich has probably been the most architecturally inspirational zoo in Europe, alongside the London Zoo. Most recently, Markus Schietsch Architekten built the elephant enclosure, a column-free large hall whose structural form resembles a turtle shell. As part of the Masterplan 2050, Zoo Zurich is moving even further away from amusing its guests. The increased communication of research findings underscores the institution's scientific aspirations, making it one of the leading zoos in Europe. In September, the zoo management under Severin Dressen (in office since 2020) presented the strategy to the public: Over the next thirty years, eleven large-scale habitats will be created. In these, different animal species will be socialized as they are in the wild. The first project to open is a large aviary for parrots in 2025. This will be followed by the Congo (2029), Sumatran rainforest, and seashore areas (both after 2030). The new habitats will all be built on the zoo's existing twenty-seven-acre site and will partially replace existing enclosures and visitor areas.

Source: Courtesy of Zoo Zurich.

in reality, analog. But we have to think carefully about which offers make sense. More animals and species or even more spectacular enclosures don't automatically mean better knowledge transfer."[3]

Pagel was until recently the president of WAZA. The network of zoological gardens is the only globally active organization of its kind that is religiously and politically independent and brings the concerns of nature conservation and environmental protection to the attention of a broad public. As such, zoological gardens have the opportunity to serve as a role model and platform for the rapidly developing ecological movements currently underway.

As mentioned, animals today are increasingly viewed as living beings with rights. This new attitude toward the relationship between humans and animals is increasingly reflected in zoo architecture. Upon closer examination, what at first sounds like a niche topic for architects turns out to be a fundamental task for our society. Building for zoo animals becomes an endless loop, in the sense that zoo architecture has always strained to create an environment for the zoo visitor that is as wild as it is for the animals. Darwin, of course, revealed that animals adapt to the environment in which they find themselves. Yet as Nigel Rothfels (this volume) writes, after more than a hundred years of experience with modern zoological gardens, the question now arises: what is actually being built—and *for whom*?

ARCHITECTURE, ACADEMICIZATION, AND WILDNESS IN THE ZOO

In order to do justice to the tasks of zoological gardens, it is important to establish building culture in zoos as an essential activity, in addition to knowledge transfer, entertainment, species preservation, and research. The Institute for Zoo Architecture, founded in 2020 at the Anhalt University of Applied Sciences in Dessau, is dedicated to this goal. The Institute of Zoo Architecture academizes a discipline that has so far only formed itself as such in practice-related projects. In addition, the typologization of buildings in zoological gardens contributes to the academicization of architecture in general. In the field of materials research, too, the subject area represents a broad field of experimentation. Material tests on special buildings, such as those that exist in zoos, enable the construction industry to evaluate them in a more sustainable way.

The Institute of Zoo Architecture is concerned with the interaction of natural habitat, artificial enclosure, and architecture. It is illusory to present animals in care as if they lived in their natural habitat. However, appropriate architecture and landscaping can help provide a comfortable environment for the animal, while presenting the zoo visitor with architecture that supports the professional and social credibility of the zoo.

Anyone walking through a zoological garden or admiring the tanks and display cases in an aquarium today will notice a new trend that started in the United States. The idea is to design the habitats in their greatest possible authenticity and to leave more room for nature and for wildness. As Ben Minteer notes in this volume, modern zoo architecture must not only be an experiential space but must also encourage people to think and, above all, act. The use of common zoological and architectural terminologies could mean that building for animals helps define a location for humans in the context of fauna. After all, the design of a zoo is always indicative of the status of the relationship between humans and animals. But thinking about zoology can also provide a new impetus for the discussion of architecture. For like nature, architecture gets by with just a few basic forms, which creativity can then vary endlessly.

NOTES

1. Lequan Chi, Christopher L. P. Wolfe, and Sultan Hameed, "Has the Gulf Stream Slowed or Shifted in the Altimetry Era?," *Geophysical Research Letters* 48 (2021), https://doi.org/10.1029/2021GL093113.
2. Author interview with Theo Pagel, October 27, 2021.
3. Author interview with Theo Pagel.

13

RECONNECTING ZOOS TO THE WILD AND RETHINKING DIGNITY IN ANIMAL CONSERVATION

JOSEPH R. MENDELSON III

The contributors to this volume present some innovative suggestions toward the stated goal of inspiring zoos to become more wild in their approach. In some cases, the recommendations relate to operational aspects of the zoo (e.g., architecture, as Natascha Meuser describes in the preceding chapter); other authors emphasize the need to embolden and better articulate the mission of a modern zoo that more centrally contributes to conservation of species, natural habitats, and reducing the negative influences of an overpopulation of humans. My fellow authors also underscore the reality that few natural areas anymore truly are wild, being far from devoid of the influences of humans. In the introduction to this volume, Ben Minteer and Harry Greene remind us that the strict preservationist ethos of nature-without-people simply is unrealistic, an artificial separation of the earth's ecosystems and the now-dominant human species. These conclusions certainly resonate with me. I teach in my courses that zoos and many national parks and reserves really are smaller or larger, respectively, models of similar functionality. I remind students that Yellowstone National Park in the United States—a crown jewel among parks globally— is, in fact, heavily managed and actually is physically fenced in some areas. I also tell them that large mammals such as bison or elk are culled for population control, especially before the reintroduction of wolves, and sometimes are vaccinated against diseases. And I point out the irony of celebrating the wildness of Yosemite National Park in California and the unspoken societal and management goal to not reintroduce grizzly bears there after they were purposefully eradicated from the entire state; all the while, the bear is featured proudly on the state flag. Wolves also have not been returned to Yosemite, although they may manage to get themselves back there one day.

Against that backdrop, I am going to propose two explicit activities that zoos can pursue that will connect them more closely with in situ conservation programs. I will end with a consideration and rejection of some loudly voiced conservation opinions from the 1980s that essentially argue that zoos are logically inconsistent with conservation programs and what is said to be wild. As Ben and Harry remind us, these sentiments have not gone away. Some conservationists argue that offering assistance to endangered species in the form of captive breeding robs the species of its dignity, thus claiming that all such zoo programs are invalid and ethically offensive. But to begin, I want to highlight an underconsidered paper on zoo programs that I believe still has some important things to teach us, especially about what a more conservation-oriented and thus "wilder" zoo could be.

William Conway, a former director of the Bronx Zoo, was a leader in zoo conceptualization and innovation in the twentieth century. An article he published in the journal *Zoo Biology* in 2011 is a pointed critique of modern zoos, outlining their general failure to create many genetically and demographically sustainable populations for even some of more popular flagship species at zoos.[1] Conway also faulted zoos for their poor track record when it came to making explicit connections between the endangered species they exhibit and in situ programs for their conservation. He contended that modern zoos should pivot away from the piecemeal collections of species that largely are driven by public popularity and should redefine themselves as "reservoirs of rare wildlife and parkland support."

Essentially, what Conway argued is that nonendangered species C has no business being in the zoo and should be replaced with endangered species D and that the latter program should have explicit ties to in situ conservation efforts for that species. He pointed major zoos toward a much more focused role that prioritizes zoo populations as crucial components of conservation programs. Zoo populations acting as ambassador animals to represent their plight in the wild are insufficient. The zoo populations have to be integrated into the conservation program. In this sense, he opined that zoos should be more similar to the International Crane Foundation (see Curt Meine's chapter in this volume) in their focus and mission. He also specifically highlighted the recently renovated lowland gorilla exhibit at the Houston Zoo for its clear linkage between the conservation challenges, the conservation efforts by this zoo, and the individuals in the exhibit, as its former director Rick Barongi has described.[2] This model also fits the recovery program for the California condor, which we will consider more deeply in what follows.

At its core, Conway's argument seems to suggest that no nonendangered species should be maintained in zoos because the space and resources are badly needed to mitigate the damage humans have done to wild populations of other species: "The survival of protected wildlife is becoming the zoo's ultimate education and conservation goal." I heartily endorse this view, but with some caveats.

As Susan Clayton and others have pointed out in this book, animals in zoos can help people become more empathetic to the plight of all animals. Effective conservation messaging can change opinions and perspectives. The goats in the petting yard, the boa constrictor brought to the schoolroom, and nonendangered species of all kinds really can highlight animals in the emerging consciousness of the public, especially perhaps in young people. So, Conway's vision is a bit too restricted for my tastes, but he is far closer to the proper direction for zoos than currently is happening. Unsurprisingly, accredited zoos across the world did not step up and meet Conway's challenge.

Inasmuch as major zoos in the United States are becoming run by CEOs with backgrounds in business and banking rather than biology or conservation, it is not surprising that the biological, conservation, and research missions at many zoos are wandering in atmospheres prioritizing revenue-generating programs and exhibits. Conservation programs are revenue-draining enterprises, of course. Leaders like Conway, Barongi, Anne Baker, Jeffrey Bonner, or George Rabb no longer are guiding the narrative, and we find a landscape of zoos dominated by animals receiving generally wonderful care but mired in nonsustainable zoo populations and vague or indirect connections to boots-on-the-ground conservation programs. These directions are what Conway admonished against in his proposal to phase nonendangered species out of zoos. I will admit that my feelings on the topic are somewhat conflicted, as I endorse a slightly less restrictive version of that goal: one prioritizing programs for threatened species at zoos but not jettisoning quite everything else. But even this clearly is not happening across zoos. So, based on twenty years as a curator and researcher at Zoo Atlanta but also coming from a background in academia, I have some observations, insights, and perspectives to share.

Larger zoos, in contrast to smaller local zoos and nature centers, often prioritize exotic animals over the native fauna. The best example of the opposite, of course, is the innovative Arizona-Sonora Desert Museum, the convening venue for *A Wilder Kingdom* (see especially the chapter by Debra Colodner et al. in this volume). This emphasis on the exotic bothers me because I feel it caters to and further encourages visitors' assumptions that all of the most interesting animals live elsewhere, for example, the tropics, and that all of the most dire conservation problems also occur elsewhere. Neither is true, of course. The southeastern United States has, by any measure, the greatest diversity of salamanders in the entire world, and they have fascinating life histories. But they are reclusive and do not make for engaging exhibits. Some of the most endangered species on the planet are endemic to North America, including the whooping crane, the Wyoming toad, the Mississippi gopher frogs, the black-footed ferret, and sadly very many native freshwater fishes, crustaceans, and mollusks (more on that later). Including endangered and nonendangered local species in zoos of all sizes is important, I believe, to remind visitors of the fascinating biodiversity in their own area. Seeing a copperhead or coyote is an important opportunity for people, especially as society stokes unfounded concerns about them.

The problem that zoos have created for themselves in focusing on exotic species is that it truly is difficult for a zoo to contribute in any tangible way to conservation efforts for those species native to the other side of the planet. Zoos have variously employed what essentially are foreign emissaries to be involved in and contribute to overseas conservation and research programs—field biologists, in other words. This is laudable, but those positions rarely survive the inevitable budget cuts that accompany economic swings; there are few such positions still active in AZA these days. So, zoos contribute in a different way, by dedicating funding to the best foreign conservation projects they can identify. Zoos praise their millions of dollars channeled to conservation, and rightly so, but as Conway and some of my fellow authors in this book point out, the contributions still represent a tiny fraction of their operating budget.

The issue with these philanthropic financial approaches to conservation is that it exacerbates the reduced wildness of zoos, as the funding provided is entirely disconnected from the funded program. Typically, no one at the zoo has visited the field site or the local staff. If they have done so, it was a one-time visit. As such, the zoo staff is disconnected from this venture. Although labor and time intensive—and exhausting—I firmly believe (as do Alison Hawthorne Deming and Gary Nabhan in this volume) that person-to-person storytelling is the best way to imbue appreciation for the plight of endangered species and habitats. If no one at the zoo has any personal familiarity with the in situ programs, the connection with the public always will be tenuous.

Rather than belabor an obvious point, that is, that zoos should be more engaged with their funded programs overseas, I will offer a more practical suggestion to better connect zoos with in situ conservation programs. Drawing from my previous points about the plight of so many endangered species in regions of the world where the larger, better funded, and accredited zoos are located, I suggest that zoos have been quite remiss in engaging these regional species centrally into their operations. There are plenty of examples of zoos doing just this, so I do not mean to imply that major zoos are not expending significant resources to protect local species.[3] I just argue that there should be so much more. Is it too much to ask for AZA accreditation to require demonstrated significant investment of space, staff time, and resources to regionally threatened species? Smaller zoos should adopt one such species, larger zoos more.

To contrast with the focus that zoos place on the exotic and spectacular, strong and direct connections with regional programs are not difficult to create. Mussels, crayfish, and unremarkable freshwater fishes are beyond the values of zoos and even many aquariums. Yet any zoo that can manage fishes or amphibians can make significant contributions to these species by doing what zoos do best: providing optimal welfare that results in the production of healthy offspring. Currently, many programs for species like this, and even for some high-profile vertebrates, are being managed in federal or state fish hatcheries or similar institutions. Why are zoos not intimately engaged in these programs? That is my provocation to the zoo world. The Saint Louis Zoo recently celebrated the release of the ten thousandth captive-reared hellbender salamander into regional waterways.[4] This zoo is wilder than many, and hellbender populations are stronger because of this ambitious and not necessarily expensive program.

Related to these points, solid engagement in regional conservation programs involving ex situ activities is an opportunity for the zoo's education and care staff to become involved, so they can experience the realities of conservation programs, become more connected to the focal species, and proceed to tell their personal compelling stories to the public. They can point to individual animals in the collection that are making genetic contributions to the wild populations. This is where, for example, African elephant programs in U.S. zoos cannot compare. U.S. zoos contribute financially to field programs for the species, but the broader disconnection is unavoidable, and no one can point to an individual elephant in a typical zoo and explain how it is helping mitigate the in situ challenges in any direct manner. When I entered the zoo profession as a complete outsider, I was surprised to learn that most of the staff on the animal, education, and veterinary teams had little or no field experience. The handful that seek out field experiences largely do so on their weekends and vacations. Fieldwork is not part of their job description.

These viewpoints serve to exacerbate the disconnect between zoos and the wild. Professional development opportunities in zoos focus on nonwild activities like workshops and training courses. Recently, at my zoo, I was championing the Caretta Project, a long-running sea turtle conservation program on the coast of Georgia. The field site is just a few hours' drive from Atlanta, and I had successfully lobbied for my zoo to provide some funding for the program. I also saw opportunities for zoo staff to intern at the site and work directly in a field conservation program. The zoo's administration was utterly uninterested in encouraging staff participation. One curator said to me, without a trace of irony: "How will this opportunity help the keepers take better care of their animals?" More zoos can get much closer to Conway's vision, without a drastic change in operations and vast financial expenditures, if they simply opened their eyes to ongoing regional conservation programs and seek to become more engaged with them.

Yet there remains a challenge to making zoos "wilder" as part of this more intensive conservation mission. It stems from the persistence of a certain kind of preservationist ideology that Ben and Harry discuss in their opening to this volume. It is an ideology that sees the work of species and wilderness protection as ultimately incompatible with the philosophy and practice of zoo conservation. The well-known story of the California condor served as a flashpoint in this clash of ideology and conservation programming in the 1980s.

Conservation efforts for the California condor feature outsize contributions of several zoos in its ongoing recovery. As Ben and Harry mention, there was a boisterous objection to that entire program as it was being developed, a part of the story that has been largely forgotten over the decades as the world watched increasing numbers of condors spread slowly across their historic ranges in western North America. The decision to capture all of the remaining twenty-three wild condors and move them into a never-before-attempted captive breeding program based in zoos was a bold one. It was quite risky, and the program developers knew they were under intense pressure to succeed. Forty years later, most would look back and say that the program has been a success, given that in 2020 there were 504 living condors, including 329 in the wild.[5] However, back in the 1980s the high-profile conservationist David Brower and his allies argued in various essays and editorials that the program was unavoidably bound to fail.[6] They rightfully were concerned about the risks of such an ambitious and unprecedented program but claimed failure to be unavoidable not because the birds would fail to breed but because the birds were in zoos. Thus, through their own ethical lens, the numbers reported in that 2020 annual report are a colossal failure. A failure not because the numbers are inadequate or because considerable expenses were involved but because condors were in zoos.

Brower and associates argued passionately for a completely hands-off conservation program to save the dwindling condor. That is, they sought to expand and better enforce protected lands around the remaining twenty-three birds, eliminate certain poisons used to control coyote populations in ranching operations, and keep all biologists well away from the birds. They viewed measuring birds, affixing visual or radio tags to them for better monitoring, and especially captive programs as ethically wrong. They then articulated a vision that I have never quite been able to absorb, and that I do not admire. Acknowledging that their hands-off approach may ultimately fail to save the species, given the dire condition of the population, they chose to identify success in the case of that reality as "extinction with dignity." The California condor, they argued, should be allowed to go extinct with dignity rather than suffer any form of direct human assistance. To them, a condor with a number tag or radio or that had been hatched in captivity had been stripped of its dignity and was not even to be considered a proper condor.

Dignity is a human construct and ultimately is a self-assessment made by an individual.[7] Assessing the dignity or lack thereof of another individual person is fraught with ethical problems and arrogance. Recall that the colonizing nations of previous centuries projected a lack of dignity on the cultures they subjugated, in part because they wore fewer clothes in their warm tropical climates. Philosophers of science tell us that species can be considered as individuals, in part because they have a beginning, an existence, a unique essence, and an end.[8] Kelsey Dayle John and Mariah ShieldChief (this volume) remind us that animals are communities comprising their own nations. In light of these points, and given the fact that we have no business assessing the dignity of anyone other than our own person, how can we legitimately project or assess the dignity of nonhuman species?

To Brower and other arch-preservationists, the condor is a symbol of the rugged landscape they roam, and the two are essentially inseparable. This view is valid on many levels, but I fail to see how it justifies avoiding conservation activities that may help save the species—activities that we now know were quite effective in steering condors toward recovery. The style and tone of their writings are imbued with phrases and imagery that seem influenced by some earlier environmental writers who conflated wildness with rugged human individualism and survivalism in remote areas. Places where a man (*sic*) can be freed from the influences of man. In embracing this testosterone-fueled idealism, they turned the condor into fetish object of their views of nature and projected their sense of dignity onto the hapless bird. Had we listened to them, the condor would certainly have gone extinct in the context of Browerian dignity.

On such purist accounts of the wild, the Herculean efforts to recover rhinos detailed by Michelle Nijhuis in this book will be a failure, regardless of eventual reproductive success, simply because the rhinos were at zoos and technology applied to assist in reproduction. Brower and allies, and contemporary bearers of such views, find no room for compromise on these issues. For me, however, the question of failure and success in these efforts is more nuanced and more poignant.

I was part of a last-minute emergency intervention by zoos to save Rabbs' fringe-limbed treefrog (figure 13.1) from extinction, an effort that ultimately failed. The species became extinct because of a fungal disease unwittingly spread to wild populations by human activities, and the efforts of three institutions ultimately failed to establish a breeding colony.[9] When I reflect on that difficult experience, I find some dignity in myself that at least we tried to help. I don't try to project dignity into the ethos of the poor frogs, just as I don't judge the dignity of the other people who worked on the project.

Relatedly, Brower-style preservationists must consider that the remaining mountain gorillas in Rwanda have no dignity because they have human bodyguards protecting them from poachers daily and because orphaned individuals are safeguarded in human-care facilities in Congo, even in the middle of a civil war.[10] I perceive a lot of dignity represented in the amazing 2014 film *Virunga*, and I'm not talking about the gorillas. The loggerhead sea turtles at the Caretta Project in Georgia emerge from natural nests protected by plastic mesh from artificially large populations of human-subsidized mesopredators like raccoons. The hatchlings emerge through the mesh and into the hands and calipers of conservation biologists who quickly tag them and let them find their own way to the ocean, but the weeks in a mesh-protected nest and the moments of measurement apparently strip each turtle of its Browerian dignity.

Let's consider what extinction with dignity looks like in the real world. The Atitlan flightless grebe, or *poc* in the regional Mayan language, was an odd species endemic to the relatively small Lake Atitlan, nestled in the chain of volcanoes along the southern coast of Guatemala. The bird was inedible and lacked gorgeous plumage, so it mostly was ignored by humans in the area for millennia. As human populations increased through the 1900s, so did the development of tourist homes and harvest of bullrushes by locals along

FIGURE 13.1. Rabbs' fringe-limbed treefrog (*Ecnomiohyla rabborum*). My colleagues and I accidentally stumbled across this previously unknown species while conducting emergency collections in anticipation of the emergence of amphibian chytrid fungus at the site. The collected frogs were to form survival-assurance colonies for a number of species of amphibians in central Panama. We formally named the new species in honor of legendary conservationists George and Mary Rabb. Two institutions in the United States and one in Panama tried to develop captive breeding programs for this species, but none were successful. The fungus arrived at the site and eradicated the only known wild population of Rabbs' frog. It became extinct when the last female died in Atlanta in 2016.

Source: Image courtesy of Brad Wilson.

the lakeshore. Introduction of North American basses, to launch a sportfishery for tourists, proved to be the death knell for the bird: the basses competed for food and perhaps consumed the baby grebes. The intrepid U.S. ornithologist Anne LaBastille and the Guatemalan conservationist Edgar Bauer raised meager funds, launched studies of the biology of the *poc*, and arranged for the government to designate protected areas.[11] Ultimately, the species was overwhelmed by the multifaceted human factors, complicated by some hydrological influences from a major earthquake to the lake and unclear influences by an ecologically similar grebe that increased in local numbers. Additionally, all conservation activities were thwarted by the ongoing genocidal civil war in Guatemala; Bauer was one of several conservationists assassinated in that era. The *poc* enjoyed extinction with dignity, never having been subjected to a captive breeding program, even if it did suffer the indignity of sometimes being captured, measured, and released and having its nests checked for eggs by LaBastille and Bauer. Barring the last bit, this evidently is what Brower and associates would prefer for species succumbing to the direct and indirect influences of human activities. Extinction with dignity.

I wish that a captive-breeding program or a rear-and-release program for chicks had been developed for the *poc*, but with ample experience in war-torn Guatemala in the 1980s, I understand why these may not have been possible. I am glad that the California condor has such a program. I hope that more zoos listen to Conway's vision and that more zoos consider my suggestions of involving themselves in regional conservation programs. I hope that people reject the notions of Brower and other wilderness ideologues and recognize the arrogance they represent. I am glad that the Saint Louis Zoo has augmented threatened wild populations of hellbender salamanders, and I sincerely doubt that those thousands of salamanders suffer delusions of indignity from their time in aquaria as they forage for crustaceans in native waters. Any disabled person will tell you that there is dignity in asking for help, and most humanitarians will argue that there is no presumption of indignity when help is offered to people in need but incapable of asking or hesitant to ask for help. Helping species in need, especially those in need because of negative human influences, is the right thing to do. We have "positive duties" with regards to conservation of wild populations, per the pragmatic preservation that Ben discussed in the introduction. Dignity of the species involved is not a relevant concept. Zoos have shown that they can be miracle workers in this realm, as in the case of the condor and many others. Let's rewild zoos and focus far more attention and resources on engaging them in what they do best. Act locally and think globally, indeed.

NOTES

1. William G. Conway, "Buying Time for Wild Animals with Zoos," *Zoo Biology* 30 (2011): 1–8.
2. Rick Barongi, "Committing to Conservation: Can Zoos and Aquariums Deliver on Their Promise?," in *The Ark and Beyond: The Evolution of Zoo and Aquarium Conservation*, ed. Ben A. Minteer, Jane Maienschein, and James P. Collins (Chicago: University of Chicago Press, 2018), 107–21.
3. "Eastern Indigo Snakes Reared at Zoo Atlanta Reintroduced to the Wild," Zoo Atlanta, July 5, 2022, https://zooatlanta.org/eastern-indigo-snakes-reared-at-zoo-atlanta-reintroduced-to-the-wild/.
4. "Historic Conservation Milestone Achieved: The 10,000th Hellbender Released Into an Ozark River by Missouri Department of Conservation and Saint Louis Zoo," Saint Louis Zoo, August 16, 2022, https://www.stlzoo.org/about/contact/pressroom/pressreleases/10000th-hellbender-released-to-the-wild.
5. U.S. Department of the Interior, *California Condor Recovery Program: 2020 Annual Population Status* (Ventura, CA: US Fish and Wildlife Service,

2020), https://www.fws.gov/media/2020-california-condor-population-status.

6. Essays by Brower and allies are compiled here: David Phillips and Hugh Nash, eds., *The Condor Question: Captive or Forever Free?* (San Francisco: Friends of the Earth, 1981).
7. Lennart Nordenfelt, "The Varieties of Dignity," *Health Care Analysis* 12 (2004): 69–81.
8. See review by David L. Hull, "Are Species Really Individuals?," *Systematic Zoology* 25 (1976): 174–91.
9. Joseph R. Mendelson III, "Shifted Baselines, Forensic Taxonomy, and Rabbs' Fringe-Limbed Treefrog: The Changing Role of Biologists in an Era of Amphibian Declines and Extinctions," *Herpetological Review* 42 (2011): 21–25. See also Joseph R. Mendelson III, "A Frog Dies in Atlanta, and a World Vanishes with it," *New York Times*, October 10, 2016, http://www.nytimes.com/2016/10/10/opinion/a-frog-dies-in-atlanta-and-a-world-vanishes-with-it.html.
10. Martha M. Robbins et al., "Extreme Conservation Leads to Recovery of the Virunga Mountain Gorillas," *PLOS One* 6 (2011): e19788.
11. Anne LaBastille, "The Giant Grebes of Atitlán. A Chronicle of Extinction," *Living Bird Quarterly* 11 (1992): 10–15.

14

SEEING THE WILD IN ZOOS BY SEEING THE HUMANS TOO

AMANDA STRONZA

Zoos have the power to shape how we understand and relate to animals. They can tell us who we are, too, we human animals; what we value, what we uphold, and how we see the world. Looking at ourselves through zoos, we can notice in particular how we frame the wild. What do we see? What do we miss? Whom do we ignore?

In this chapter, I explore how zoos can widen the frame of what we see and understand. I suggest that with thoughtful attention to stories, and who gets to tell them, zoos can build greater awareness of how animals are connected with humans and how conservation efforts must support strategies for coexistence. My definition of zoos is expansive. I am recognizing that zoos, sanctuaries, and parks fall on a continuum of spaces that are not-quite-wild. They all restrict the movement of animals, however greatly or minimally; they all involve managing animals, either as individuals or populations; and they all invite humans to visit, see, and learn. In some ways, the places where humans live close to wild animals are "wilder" than the set-apart spaces of zoos and parks. Perhaps "wildness" is relative rather than absolute.

I will suggest that the "wild" therefore includes humans and that "wilder" zoos and parks can embrace new stories, wider angles, and help us value landscapes that are both social and ecological. Zoos can also inspire visitors to learn and care about the relationships between people and other animals. I will focus on elephants, in particular, asking how zoos and parks can support wild elephants and their families while also building understanding of how humans relate to elephants and how elephant stories overlap with human stories. If zoos are our mirrors, they can also be places to magnify our best intentions, our biggest goals of honoring, protecting, and ultimately coexisting with wild animals.

THE KNOTTIEST QUESTIONS

Discussions about the pros and cons of zoos take us to the knottiest, most impassioned questions about animal rights, animal welfare, conservation, extinction, ethics, privilege, choice, education, crisis, and triage. For many, zoos are essential for conservation. Proponents note that zoos have safeguarded reservoir populations of endangered species while also funding conservation projects in the field and leading fundamental research on animal behavior and biology.[1] The American bison and the California condor are two species whose wild populations benefited from the captive breeding and conservation efforts of zoos.[2] And in her chapter that follows, Michelle Nijhuis describes the heroic efforts underway at zoos like San Diego to recover the northern white rhino, a subspecies on the ecological ropes. Zoos have a record, too, of building feelings of stewardship and love for animals, especially among North American children and adults who otherwise never get to see capybaras, servals, rhesus macaques, or countless other creatures.[3]

Despite these considerations, some critics maintain that keeping animals in captivity is misguided and wrong. They say wild animals should never be kept "like prisoners," distressed and bored in spaces that restrict their movements and behaviors as well as their spirit and autonomy.[4] For every bison and condor story, there are accounts of animals in zoos with phobias, depression, pacing, and other disorders that have no analogues in the wild.[5] Still, some zoo advocates suggest that confinement can be less stressful, at least acutely so, than the wild. In zoos and other managed spaces, animals are not as vulnerable to the stresses of climatic extremes, predators, parasites, injuries, starvation, or other hardships.[6]

Yet zoos are not monolithic. Some confine animals in relatively small spaces; others offer freedom to roam. Some treat animals as objects of entertainment; others care for animals as family. Some offer misinformation or virtually no information; others specialize in entire K–12 education programs. There are bad zoos and better ones. A "bad" zoo makes animals work for it; a "good" zoo works for animals.[7] Perhaps the "best" zoo can work for animals and also the humans who live with them, in the wild and outside of parks.

The debates about zoos are especially resonant for certain "more-like-human" species, such as great apes, cetaceans, and elephants.[8] Elephants have long been popular at zoos.[9] For most people, a zoo is the only place they may ever see an elephant. Yet zoos and even many parks lack space that is big enough, rich enough, or stimulating enough to sustain large populations. Elephants need space to let their minds roam as much as their bodies. Ultimately, the stress of being in captivity takes a toll. The average life span of a captive elephant is seventeen years, at least three decades shorter than in the wild.[10] Though the standard set by the Association of Zoos and Aquariums (AZA) for one elephant is 1,800 square feet,[11] elephants are in constant motion, roaming tens of kilometers each day, hundreds of kilometers seasonally,[12] sometimes traveling especially far so that multiple family groups can reunite and bond in large clans.[13] Even if it were possible to provide the right space, social support, and care for elephants, most zoos, and even most parks, lack the resources. Elephants cannot really be in the room.

BUILDING CONNECTIONS, REINFORCING SEPARATION

So far, I have likened zoos to parks, noting how both limit the movements of animals, sometimes in small enclosures, sometimes in large expanses, and how both invite humans as visitors. What zoos and parks also share is the power to make people feel connected with animals. This is true even though most erect barriers, and most reinforce (even if unintentionally) the idea that humans are exceptional. This is a conundrum: how can zoos help people feel connected with animals, either the ones exhibited or all animals, even while constructing physical and conceptual divides? How can zoos deepen our understanding of animals as sentient beings, maybe even as kin, even while separating them from us?

Being in the presence of wild animals—sensing, hearing, smelling, watching how they behave and interact with one another—can be so powerful. Nothing compares with seeing an elephant in person (figure 14.1). The chance to look into the eyes of an enormous and sentient being, so different and similar to humans, is profound. These experiences build sensitivity and empathy as well as feelings of connection, care, and stewardship. Yet some of what we learn in zoos and parks is inaccurate or at least only partially accurate in terms of showing what animals are, who they are, and how they behave in the wild.

By confining animals to exhibits, however large and enriched they may be, zoos and parks can perpetuate the myth of nature as separate from culture, the belief that wilderness is a place free from

FIGURE 14.1. Few things are more stirring than meeting an elephant in person. We can learn so much just by being in their presence—hearing them rumble and vocalize with one another, watching how they express themselves with their whole bodies, and seeing the light in their eyes. Zoos have the power to share this and to make people feel connected.

Source: Photo by the author.

human history and experience. Zoos and parks can further entrench a human-nature divide rather than illuminate and celebrate the many cultural, historical, and evolutionary bonds people and animals share. Even the new, "utopic" zoos hide humans rather than include them in the landscape. But this is a problem because human culture is part of the wild.[14] Zoos and parks that are off-limits to humans except as tourists mask this reality, whether the cages are visible or veiled. Zoos can make it seem like wild animals exist in realms of the pristine, what we imagine to be the wild, even if we know it's a wild that is either held up, as in exhibited, or held back, as in enclosed by fences.

Even national parks tell stories of the wild that are false or, at least, incomplete. Though protected landscapes may appear as pristine, untrammeled wildernesses, they are also cultural spaces that have been modified and shaped by indigenous peoples through generations of harvesting, planting, burning, and hunting and later by resource managers, landscape architects, tourism companies, and recreation planners.[15] Many parks in Africa that manage and sustain elephants are modified by artificial water holes, veterinary cordon fences, and safari concessions. We must question the wild, even in the seemingly wildest places.

Zoos separate people from other animals figuratively as well. They can deepen our perception of difference or reinforce the belief in a hierarchy of humans above other species, further instilling a sense that we are exceptional, that we are the only animals who don't belong in captivity, unless, of course, we have committed a crime. Only the other animals, the ones not like us, are behind bars. Who is gazing and who is being gazed upon? Do animals have a right to deflect, a way to hide? No matter how clever or concealed the confinement may be, zoos and parks might not escape what Ben Minteer called a "moral friction": humans are "both co-inhabitants with other animals and (increasingly) creators of their worlds, including those beyond the zoo walls."[16]

Historically, even well-funded zoos, like the Bronx Zoo, had small enclosures for large animals.[17] Some contemporary zoos still do. These tend to be the underfunded ones or the ones tied with circuses, theme parks, or side-of-the road shows. The problem with these kinds of zoos, the ones that confine animals primarily for profit and entertainment, is that they deepen rather than challenge the notion of animals as Other, as objects to be seen rather than as beings with the will and freedom to interact with humans (or not) on their own terms. These kinds of zoos give license to turn off our feelings about the feelings of other animals. I visited a zoo in the Colombian

Amazon that held a manatee in a tank so small she could not turn her body. She just lay there, in stillness, still alive, but only functionally. In Indonesia, I visited a zoo that held macaques in portable cages in the midday sun. People could approach the cages and throw food and trash at the monkeys. I believe the cages enabled people to see the monkeys in a certain way, as objects, and gave license to disregard, to mock rather than care. One study found visitors to a zoo were concerned about the well-being of animals and perceived the animals as distressed and unhappy but nevertheless valued the experience.[18] Despite the tension within themselves, despite the evidence of tension in the animals, the visitors stayed, and they watched.

Physical and conceptual structures that reinforce separation and exceptionalism in zoos can preclude us from feeling empathy and connection with other animals. They can also prevent us from understanding how wild animals live in shared landscapes with humans and why we need to support on-the-ground efforts that protect wildlife while also benefiting humans.

WILDER ZOOS AND PARKS

Zoos are being reimagined and reconceived. This is not a new trend; zoos have been changing for decades, reforming practically since they were formed. Changes include replacing bars and concrete with moats and grass; making enclosures more accommodating or at least less stressful; improving diets and veterinary care; and emphasizing conservation, research, and education over entertainment. In some cases, this has meant flipping the paradigm, confining humans rather than the other animals and reversing the direction of who is gazing.[19] The changes can make zoo spaces appear more natural or wilder, even as they continue to reflect human ideals of what wild nature is. The reforms have tended to be architectural rather than foundational—tweaks in design rather than fundamental shifts in the subject featured or the story told.

As zoos have reformed to be wilder, so too have parks expanded beyond fences. Conservation efforts are increasingly focused on landscapes that are both social and ecological, giving attention to wildlife in places where people live and work, beyond the boundaries of protected areas.

What do "wilder zoos," or places with a "heightened embrace of naturalism," imply for elephants? At least two things can change: the story told and the conservation goal. As zoos move toward a wilding of spaces for animals, expanding the perimeter and purpose of their exhibits, and as parks do the same, taking into account the human lives and settlements beyond park boundaries, we can reframe what is worthy of inclusion. We can tell a more honest story. We can include the social sides of wild places, the human, cultural, and historical connections people share with wild animals. Alison Deming (this volume) writes of how including the arts in wilder zoos allow for storying animal lives, offering complements and counterpoints to species data.

This kind of change can be especially meaningful for elephants. Elephant conservation in the field, in the wild, is necessarily integrative, focused on the ways in which people and elephants interact and supporting the livelihoods and needs of humans while also protecting the habitats and needs of elephants. Human-inclusive stories can reflect more accurately what elephant conservation looks like in the field, in lands outside of parks, where so many elephants live. For elephants, "in the wild" means "with humans." A "wilder zoo" for elephants implies a wider angle on the stories we tell.

Something as simple but profound as reframing interpretive materials and stories about elephants can help build connections rather than reinforce separations. Even in zoos where elephants are confined or in parks and sanctuaries where their spaces are limited or fenced off from humans, expanded stories can build support for conservation efforts that focus not just on protecting elephants but on alleviating human-elephant conflict and enabling coexistence.

A WILDER, WIDER ANGLE ON ELEPHANTS

A "wilder" zoo or park is one that takes a wider angle on elephants and helps visitors understand human-elephant conflict as one of the greatest threats to elephants and to the people who live with elephants throughout rural communities in Africa and Asia. No single park in the world, much less any zoo, is big enough to sustain large populations of elephants. In some cases, where elephants have been translocated from open, rural spaces to parks, they either suffer from a lack of resources, or they break out of park boundaries and face conflict with people. The elephant range necessarily overlaps with human settlements.

It is in these spaces of competition over resources, places that have been called "elephant landscapes," that human-elephant conflict is a challenge. For people who live with elephants, "conflict" comes in the form of elephants damaging or eating crops, sometimes depleting an entire year's supply of food in one night. It also comes when elephants destroy property or fences and, most devastatingly, when elephants charge at, threaten, or kill people. For elephants, "conflict" comes when people clear habitat or block access to water and, most devastatingly, when people chase, harass, or kill.

To know elephants in the wild is to know it is impossible to talk about them without seeing how they live with people. The fact that elephants throughout their range have always lived with people is both a challenge and an opportunity. Stories of coexistence intersect with stories of conflict. The only way to ensure elephants can survive in the wild is to embrace rather than resist the fact that they must survive in vast social-ecological systems, with humans. Elephants must live among us or not at all.

Much of the media attention on elephants in recent decades has been on the dramatic loss of African elephants to the ivory trade. By one estimate, 100,000 elephants were killed illegally for their tusks in three years. But if the ivory trade has been a meteorite striking elephant populations, human-elephant conflict, or HEC, has been death by a thousand cuts. Even as poaching has begun to taper, HEC looms. Conflict is a threat to all species of elephants throughout their ranges in Africa and Asia. It is the primary concern for the IUCN Elephant Specialist Group and the Elephant Crisis Fund.[20] For Botswana, the country with the largest population of elephants on the planet, HEC is a chief policy concern and an urgent and daily challenge for people in rural villages. It is at the very heart of overlapping, vexing challenges of poverty, food insecurity, social inequality, habitat loss, and climate change.

Despite the pervasiveness of HEC, zoos and parks share little of that story. We rarely see or hear stories of HEC in interpretative materials or signs or talks. We hear even less from people who can convey in raw and powerful terms what HEC feels like, who know first-hand what it's like to stand vigil through the night in a watch hut beside a field of millet and maize, hoping elephants won't come and having only a string of tin cans as defense in case they do. These are the stories that need to be told, especially in creative collaboration with people who actually live with elephants.

Zoos and parks have an opportunity to bring local voices, knowledge, and experiences into the interpretation of what elephants are, who they are, and what they represent to people who live with them. Like the parable of the blind men touching different parts of an elephant, all having the conviction of being right, we perceive elephants differently depending on where we stand. Rural people who live close to elephants know a different beast from policy makers who live in faraway cities. Animal lovers who see gentle giants in zoos may have no comprehension of how dangerous elephants can be.

The stories from local voices are not all negative. There are stories of reverence, kinship, and connection, too (see fellow authors Harry Greene, Curt Meine, and Gary Nabhan in this volume). There are stories of how people are learning to share space with elephants, to ensure that movement corridors are respected and protected; that farmers' fields are clustered and guarded from elephants in innovative, nonlethal ways; that people are developing microenterprises that build on their shared life with elephants; that people have deep knowledge of elephant behavior, movement, and culture, and that this knowledge is often expressed in creative ways through dance, poetry, painting, and basketry. There are many stories to tell.

Changing interpretive stories about elephants can help build support for elephant conservation and also empathy for people who bear the cost of living with elephants in wild spaces, which, perhaps to the surprise of visitors, includes villages and farms. A wider angle on elephants will help zoo and park visitors understand that all forms of confinement, whether 750-foot enclosures, three-hundred-acre sanctuaries, or landscape-size parks, are artificial. All of them uphold a myth of wild nature, one that is separate from the lives of people. It is a false story because elephants in the wild, in the truly wild, live with people.

NEW STORIES, NEW CONNECTIONS

At the root of many conservation challenges is an estrangement between people and nature, a disconnect tied with a belief that humans are different from other animals, exceptional. Zoos and parks can reinforce the idea that humans stand apart by literally keeping them apart, erecting fences and plexiglass, allowing people to get close enough to gaze but not close enough to understand or really feel connections and understand that people are part of nature.

Looking forward, "wilder" zoos and parks can bring down at least the conceptual barriers and tell more expansive and inclusive stories about animals, offering wider views of the wild, ones that include humans. As Carl Safina has suggested, a goal of zoos can be "not to separate humans from other animals, but to entangle all humans in nonhuman lives."[21] Rangers, guides, interpreters, and docents in wilder zoos and parks can help people perceive other animals as connected with us, rather than apart from us or beneath us; as fellow beings rather than as objects, as "us" rather than "them."

Wilder zoos with wider views can shine light on how conservation needs to happen in places where people live, beyond exhibitions and even beyond the fences of parks, in landscapes that are both social and ecological. Such places could support research, too, expanding beyond the biological and behavioral research of zoos, paying attention as well to broader social-ecological systems and histories and to the complexity of relationships between humans and other animals through the lenses of the social sciences, arts, and humanities.

Though wilder zoos and parks would be managed, perhaps they could support behaviors, preferences, and interactions that appear in less confined spaces. For example, in the wild, in shared social-ecological landscapes, elephants try to avoid people as much as possible. People do the same with elephants. Could a wilder zoo or park allow for such mutual avoidance? Also, in the wild, elephants often adhere to certain movement corridors, pathways that are chosen and remembered by matriarchs and their families for generations, regardless of park boundaries. These are too expansive for zoos, but wilder parks could ensure that corridors are honored, protected, and included in the stories visitors hear.

Wilder zoos are an opportunity as well to celebrate the ways in which indigenous and rural peoples of the world have coevolved with other species, have maintained, shared, and entrusted traditional knowledge about wildlife that is valuable and essential for conservation. It is a chance to give visitors a way to help beyond giving money to captive animals, to channel their concern and goodwill to people who are in many ways struggling to coexist with wildlife.

I am not suggesting a return to the era when indigenous people were on display in museums and fairs, featured as "savages," "like animals," objectified and voiceless. Instead, my recommendation is that we hear and learn more from people who have stories to tell, varied, inspiring, challenging, and troubling stories, as well as have so much knowledge about what it's like to coexist "in the wild" with elephants and other wild animals.

CONCLUSION

Our efforts to make zoos and parks "more like the wild" also represent opportunities to make them "more human." When we give thoughtful attention to the social dimensions of the "wild," we can reflect on how wild animals share life histories and life stories with humans. This wider view can help people see, understand, and value elephants differently, inspiring visitors with powerful stories not just of elephants but also of the complex relationships between elephants and humans. This perspective can also lead to meaningful support for rural people everywhere who live side by side with elephants and other wild animals and who serve as stewards and knowledge holders. Our shared, sustainable future, a world where coexistence is possible beyond zoos and parks, depends on these feelings of connection and empathy.

NOTES

I am indebted to Dr. Anna Songhurst, Dr. Graham McCulloch, and countless people and elephants in the villages of the Okavango Panhandle in Botswana who taught me to see the connections and stories of coexistence between people and elephants.

1. Oliver A. Ryder and Anna T. C. Feistner, "Research in Zoos: A Growth Area in Conservation," *Biodiversity and Conservation* 4 (1995): 671–77, https://doi.org/10.1007/BF00222522.
2. Carl Safina, "Where Are Zoos Going—or Are They Gone?," *Journal of Applied Animal Welfare Science* 21 (2018): 4–11, https://doi.org/10.1080/10888705.2018.1515015.
3. Benjamin Wallace-Wells, "The Case for the End of the Modern Zoo," *New York* Magazine, July 11, 2014, https://nymag.com/intelligencer/2014/07/case-for-the-end-of-the-modern-zoo.html; Tim Zimmerman, "The Case for Closing Zoos," *Outside*, February 13, 2015, https://www.outsideonline.com/culture/opinion/case-closing-zoos/.
4. Lynn Griner, *Pathology of Zoo Animals: A Review of Necropsies Conducted Over a Fourteen-Year Period at the San Diego Zoo and San Diego Wild Animal Park* (San Diego: Zoological Society of San Diego, 1983); Barbara King, *Animals' Best Friends: Putting Compassion to Work* (Chicago: University of Chicago Press, 2021).
5. Alex Halberstad, "Zoo Animals and Their Discontents," *New York Times Magazine*, July 3, 2014, https://www.nytimes.com/2014/07/06/magazine/zoo-animals-and-their-discontents.html.
6. Nigel Rothfels, *Savages and Beasts: The Birth of the Modern Zoo* (Baltimore, MD: Johns Hopkins University Press, 2002), 199; Dave Hone, "Why Zoos Are Good," *Guardian*, August 19, 2014, https://www.theguardian.com/science/lost-worlds/2014/aug/19/why-zoos-are-good.
7. Safina, "Where Are Zoos Going—or Are They Gone?"
8. Wallace-Wells, "The Case for the End of the Modern Zoo."

9. Jeffrey P. Cohn, "Do Elephants Belong in Zoos?," *BioScience*, 56, no. 9 (2006): 714–17, https://doi.org/10.1641/0006-3568(2006)56[714:DEBIZ]2.0.CO;2.
10. Zimmerman, "The Case for Closing Zoos."
11. Cohn, "Do Elephants Belong in Zoos?"
12. Ian Sample, "Stress and Lack of Exercise Are Killing Elephants, Zoo Warned," *Guardian*, December 11, 2008, https://www.theguardian.com/science/2008/dec/12/elephants-animal-welfare.
13. Cynthia J. Moss, *Elephant Memories: Thirteen Years in the Life of an Elephant Family* (Chicago: University of Chicago Press, 1988).
14. Ben A. Minteer, "The Real Zootopia," *Slate*, March 3, 2016, https://slate.com/technology/2016/03/the-real-zootopia-in-denmark-has-experts-divided-on-the-state-of-the-zoo.html.
15. Minteer, "The Real Zootopia."
16. Minteer, "The Real Zootopia."
17. Safina, "Where Are Zoos Going—or Are They Gone?"
18. Prokopis A. Chirstou and Elena S. Nikiforou, "Tourists' Perceptions of Non-Human Species in Zoos: An Animal Rights Perspective," *International Journal of Tourism Research* 23, no. 4 (2021): 690–700, https://doi.org/10.1002/jtr.2435.
19. Wallace-Wells, "The Case for the End of the Modern Zoo"; Rothfels, *Savages and Beasts*; Halberstad, "Zoo Animals and Their Discontents"; J. Weston Phippen, "Do We Need Zoos?," *Atlantic*, June 2, 2016, https://www.theatlantic.com/news/archive/2016/06/harambe-zoo/485084/; Michael Gross, "Can Zoos Offer More Than Entertainment?," *Current Biology* 25, no. 10 (2015): R391–94; Becky Quintal, "BIG Unveils Design for 'Zootopia' in Denmark," *ArchDaily*, July 29, 2014, https://www.archdaily.com/532248/big-unveils-design-for-zootopia-in-denmark; Ben A. Minteer and Harry Greene, "Zoos and the Wild: A Reconsideration," this volume.
20. George Wittemyer et al., "Illegal Killing for Ivory Drives Global Decline in African Elephants," *PNAS* 111, no. 36 (2014): 13117–21, https://doi.org/10.1073/pnas.1403984111; Elephant Crisis Fund, https://www.elephantcrisisfund.org/.
21. Safina, "Where Are Zoos Going—or Are They Gone?"

15

THE ONCE AND FUTURE RHINO

MICHELLE NIJHUIS

On a sunny fall day in southern California, deep within the San Diego Zoo Safari Park, a two-and-a-half-month-old baby rhinoceros named Edward was gamboling around his mother.[1] One hundred and forty-eight pounds at birth, his weight had since tripled and then some, and two tiny nubs of horn were rising on his snout. When his mother approached the gate of their shared enclosure, where a bucket of hay was on offer, Edward bustled over, too. He poked his huge, dusty jaw through the steel bars, snorting as he submitted to affectionate scratches from human hands.

Edward is a singular animal, famous since birth for the circumstances of his conception: he developed from an egg that was fertilized in the laboratory and implanted in his mother's womb. Though his subspecies, the southern white rhino, is not immediately threatened with extinction, his birth and good health are small steps on the long, uncertain road toward the resurrection of the northern white rhino—a central African subspecies that poaching and habitat destruction has reduced to just two surviving individuals, both female.

Researchers at the zoo and elsewhere hope to "rewind" the functional extinction of the northern white rhino by using genetic techniques to turn frozen skin cells into stem cells and then into viable sperm and eggs.[2] If they are successful in creating northern white rhino embryos, they will implant them in the uteruses of surrogate southern white rhino mothers, raise the resulting northern white rhinos in captivity, and eventually—if the risk of poaching can be reduced—introduce the northern white rhinos to their native range.

But none of these techniques has been fully developed for northern white rhinos—or any rhinos, for that matter. The genetics research remains in its early stages. The standard method of in vitro fertilization requires too large a fraction of the small existing reserve of frozen northern white rhino sperm, so an alternative method has been adopted. Embryo implantation requires a specially designed robotic catheter able to navigate the rhino cervix, whose canal is so tight and twisted that reproductive scientists describe it as "tortuous."

In early 2018, veterinary technicians successfully implanted a southern white rhino embryo into Edward's mother's uterus, and after a sixteen-month pregnancy and a thirty-minute labor, she gave birth to Edward on July 28, 2019, demonstrating that the zoo's fertilization and implantation methods could be used to produce a healthy baby rhino. Four months later, another rhino gave birth to a female calf named Future, also conceived through artificial insemination. Step by expensive, nerve-racking step, the rewinding attempt continues.

For zoological parks to become wilder, in any of the senses of "wild" discussed in this volume, they must affirm and expand their contributions to a wilder world—a world, that is, in which more species are not only safe from extinction but protected in abundance in their own habitats, however those habitats are defined. Can the northern white rhino project and other highly experimental ex situ efforts make such a contribution?[3]

FIGURE 15.1. Edward and his mother at the San Diego Safari Park in October 2019.

Source: Photo by the author.

Zoos first became actively involved in conservation when William Temple Hornaday raised a herd of Oklahoma-bound bison in the Bronx at the start of the twentieth century. Animal reintroductions attempted since then have frequently failed,[4] but in a few key instances, captive-bred animals have helped reestablish extinct populations and augment those threatened by inbreeding or disease.

The San Diego Zoo has a record of what Oliver Ryder, director of conservation genetics for the zoo's Wildlife Alliance, calls "hopeful interventions." Its researchers had a central role in the captive breeding and release of the California condor in the early 1990s and are global pioneers in high-technology husbandry. Its so-called Frozen Zoo, founded in 1975, contains cryopreserved tissue, sperm, and eggs from more than one thousand vertebrate species, and among its holdings are cell lines from twelve individual northern white rhinos. The largest collection of its kind, the Frozen Zoo enables ongoing research on advanced reproductive techniques and other methods of rescuing tiny populations from extinction.

As excited and pleased as Ryder and his colleagues have been by Edward's birth and subsequent advances in their research, they know from experience that years, perhaps decades, of costly work remain—and that failure is far more likely than success. If they do succeed in the laboratory and the rhino enclosure, they know that they will have only reached the starting line of traditional species conservation. All the complexities of controlling poaching and protecting habitat—of working with governments and community organizations in central Africa to provide a reestablished rhino population with the food, water, space, and security it needs to reproduce, adapt, and persist—will still lie ahead.

They also know that navigating these complexities, whether ex situ, in situ, or somewhere along the spectrum between, is intrinsic to their vision for the project. In late 2019, Ryder and his Wildlife Alliance colleagues convened a multidisciplinary group to discuss the ethics and social dynamics of the northern white rhino initiative. In a subsequent paper, Ryder and several coauthors argued that while the initiative offered important near-term opportunities to develop new assisted-reproduction techniques and foster a connection between zoo visitors and the northern white rhino, researchers were ethically obligated to extend its benefits beyond "affluent researchers, affluent organizations, and affluent tourists."[5]

Without the goal of returning northern white rhinos to their historical range, the authors observed, the project risked creating a moral hazard, signaling to zoo visitors and the broader public that a species or subspecies could be rescued through technology alone. "It is the audacity and ambition of the project as conceived in its fullest sense that provides its meaning, narrative, and symbolic value," they wrote. Proposing to build on the integrated One Plan approach to species conservation developed by the IUCN,[6] they envisioned a project whose contributions will be scientific, educational, social, and ecological—a "game-changing species conservation success story" with symbolic and practical value to a wilder world.

Realizing this vision for the northern white rhino project and other captive-breeding initiatives will require the disruption of long-established patterns, beginning with the typical presentation of such work to the public. Zoos regularly promote their conservation initiatives in order to justify the continued existence of zoos and rally donor support, especially in the face of increased public concern about species extinction and individual animal welfare. Captive breeding and reintroduction efforts, with their dedicated researchers and often adorable charges, are a particularly rich source of stories that appeal to journalists and media consumers, reflect positively on zoos, and inspire zoo visitors.

After the San Diego Zoo communications staff energetically publicized Edward's birth in 2019, the event made national news and was widely celebrated in San Diego. In press coverage, the project's larger vision was mentioned only in passing, if at all, and *People* magazine's report on Edward's birth was included in its "Pets" section. While the media bears much of the responsibility for this lack of context, communications staff could more actively challenge it through their messaging. The most-used promotional photo of Edward, for instance, was typical of the "celebrity photos" described earlier in this volume by Alison Hawthorne Deming. Instead of showing the young rhino in relationship with his mother (who, after all, had done most of the work) or the analogues of rhino habitat available at the zoo, the photo showed Edward alone, running toward the camera against a nondescript background.

Today, an illustrated sign outside the rhino breeding facility confidently depicts the assisted reproduction process for the northern white rhino as a three-step recipe, the third being "Successful Reproduction." The language is present tense, not conditional, effectively erasing the laborious cycle of trial and error required to develop any new genetic or reproductive technique. The goal of reestablishing a northern white rhino population in central Africa is not mentioned, but thanks to science, the sign implies, the rescue of the subspecies is already well underway.

To a public bombarded with news of human sins against the rest of life, such images and stories are surely a welcome change. As Ryder has written, "'Science can help' is a hopeful and motivating message." The use of simplified visuals and optimistic language is understandable, given the heavy competition for media and visitor attention, and any advance toward rescuing a taxon, no matter how contingent, is worth trumpeting. But by eliding the risks, uncertainties, and ethical questions that inevitably accompany the development of new reproductive technologies, conventional zoo messaging risks further distorting public perceptions of conservation. Those of us who care about the persistence of other species are, after all, ripe for misleading: given the gravity of the threats and the complexity of systemic solutions, it's tempting to invest hope in technological fixes—to rush toward moral hazard, in other words, unless we are explicitly warned away. The sight of an energetic young rhino, clearly thriving at the zoo, only increases the temptation. Wouldn't it be wonderful if technology could allow us to bypass the long, politically fraught struggle to protect species and their habitats and simply . . . make more animals?

While it's impossible to measure the effects of zoos' institutional messaging, or for that matter untangle its causes from its effects, my experience as a journalist covering conservation issues is that in general, public perceptions of conservation mirror its standard presentation in and by zoos. Even individuals passionately interested in conservation often exaggerate the role of ex situ conservation and the rescue of extraordinarily rare species. Very few acknowledge the reality that these intensive measures are desperate last resorts and that the larger aim of conservation is to ensure they never become necessary.

This skewed view of conservation may help explain the recent fascination with "de-extinction" both in the press and in scholarly circles. While the as-yet-theoretical creation of hybrid individuals using DNA from extinct and extant species may one day have conservation applications, it will never be capable of reversing extinction. But its popular nickname, combined with the widespread sense that "de-extinction" is the central goal of conservation, has fired the public imagination and flooded its supporters with attention and funding. In September 2021, a bioscience startup called Colossal announced that it had raised $15 million to fund its attempt to create a hybrid embryo containing DNA from an elephant and a woolly mammoth. Though this research is not necessarily competing directly with conservation efforts for funding, its budget makes for a startling contrast with those of most field conservationists.

For zoos to publicly express the kind of humility and uncertainty that many zoo researchers readily express in conversation runs counter to long-established habit, and it is, admittedly, hazardous: acknowledging the many unknowns of captive breeding efforts muddies the narrative and may well create doubts in the minds of visitors and donors. Emphasizing that captive breeding is, in most cases, a response to an already tragic situation—that reproductive science researchers who specialize in endangered species are more like overworked emergency responders than bold explorers—darkens the sunny story of scientific progress. But by avoiding these truths, zoos pass up a chance to more thoroughly educate their visitors. They also risk creating the kind of complacency that Mark Vermeij, a marine biologist on the Caribbean island of Curaçao who is part of an international effort to raise endangered coral larvae in captivity, hears all too often. "People have approached us and said, 'Ah, that's nice, because now the Great Barrier Reef is fine,'" he told me. "And it's like, 'What on *earth* are you fucking talking about?'"

Historically, zoos have struggled to integrate ex situ efforts with their in situ counterparts, given time and funding constraints, institutional inertia, and the challenges of communicating across disciplines, cultures, and continents. For similar reasons, implementation of the One Plan approach has been uneven. Plans for species management and collection have been and continue to be devised without input from the field researchers most familiar with the conservation needs of the species in question. Zoo partnerships with in situ conservation organizations, while increasingly common, are not well coordinated within the zoo community, and financial contributions by zoos to field conservation efforts represent a tiny percentage of zoo budgets.[7]

The current in situ conservation strategy for the northern white rhino could hardly be simpler, or starker: the two remaining individuals live in Kenya's Ol Pejeta Conservancy, outside their subspecies' historic range, and are under constant heavy guard given the ongoing threat of poaching. Since it could well take decades for rhino researchers to establish a captive population of northern white rhinos—if they can establish one at all—developing in situ protections for a free-roaming population might seem premature. But securing and sustaining adequate rhino habitat is not as simple as building higher walls and hiring more guards. Like ex situ work, it is an uncertain process that can require decades of patience.

Powdered rhino horn, though discredited as a traditional medicine, still fetches high prices in Asia as a status symbol, and continuing demand has created an epidemic of poaching in southern and central Africa. Many parks and reserves have armed their rangers, leading to a low-level war between rangers and poachers with heavy casualties on both sides—and, so far, little reduction in rhino poaching. One bright spot is northern Namibia, where in the 1980s local communities began hiring residents—including some former poachers—to serve as unarmed game guards. The guards, who knew the terrain and were well informed by watchful neighbors, began to intercept and warn off would-be poachers of rhinos, elephants, and other species. The program was so successful that in the 1990s it became the basis of a nationwide network of community conservancies, self-governing entities that monitor and manage local wildlife for both human benefit and long-term sustainability. When a new wave of poaching hit the area in the 2010s, the game guards worked with Save the Rhino International to expand their ranks in response. The region's population of black rhinos, which nearly went extinct in the 1980s, has recovered and is now among the healthiest on the continent.

The success of the Namibian conservancies and of other community-led conservation initiatives worldwide depends largely on the strength of the relationships among their participants, and those relationships—which include not only community members but government officials, professional conservationists from near and far, and many others—can take a generation to build. Over time, community conservancies have shifted some of the practical benefits of conservation toward local people, eased some of the burdens, and gradually rebuilt some traditional conservation practices, many of which were disrupted by colonization. While the San Diego Zoo currently supports the rhino-conservation work of the Northern Rangelands Trust in Kenya and the International Rhino Foundation in South Africa, Ryder and his colleagues envision direct engagement with human communities in places once inhabited by the northern white rhino—"a comprehensive, collaborative project to use the promise of rhino return to foster ecological protection as well as social improvement and empowerment." Such collaborations can and should start sooner rather than later. By proactively establishing relationships with community-led conservation organizations in future in situ habitat, zoos can begin to lay the groundwork for successful reintroductions, supporting the conservation of resident species along the way. Even if the captive breeding project in question never comes to fruition, these sustained partnerships can still advance conservation and can serve to educate zoo visitors about in situ realities.

Long-term, substantive integration of ex situ and in situ conservation creates another opportunity for zoos to contribute to a wilder world. One of the great accomplishments of the modern conservation movement over its century-plus lifespan has been to forge new connections between humans and other species, including species that some of us may never see roam free but whose fate we nevertheless affect. Worldwide, zoos have helped inspire this concern for distant species, and they should continue to do so. By working in partnership with the human residents of potential in situ habitats, however, zoos can begin to correct one of the most damaging misjudgments of the conservation movement, which is that the people who live most closely alongside other species are not capable of maintaining them. As Kelsey Dayle John and Reva Mariah Shield-Chief write in this volume, modern conservationists have for too long ignored the Indigenous and other traditional conservation practices that have helped people coexist with other species for millennia, instead focusing on creating parks and reserves—and fences and cages—designed to protect species from people. While boundaries are often necessary in conservation, no barrier is inviolable, and not even the largest reserve network can encompass all the habitats that self-sustaining populations require. To conserve and restore healthy populations, conservationists must recognize and communicate to their supporters that people of all walks of life are eminently capable of coexisting with other species.

Zoo partnerships with community-led conservancies in in situ habitats can demonstrate this capability to zoo visitors, and zoos can draw analogies between distant attempts at coexistence and those needed closer to home. Zoos could guide visitors through areas of local native habitat, for instance, and ask them to consider how the species they live alongside are accommodated—and not accommodated—by their households, neighborhoods, and municipalities. Zoos could highlight their own efforts at coexistence with the free-roaming animals in and around their properties. In these and other ways, zoos could do much more than excite visitor sympathy for individual captive animals. By instilling a sense of ecosystem membership in their visitors, they could foster more enduring emotional and cognitive connections to our fellow species and a deeper understanding of the importance of protecting and restoring healthy populations.[8] In a wilder zoo, visitors would be consistently reminded that even the rarest, most intensively managed animals are part of the wilder world we all share.

NOTES

1. Unless otherwise noted, material in this chapter is based on original reporting and adapted from my book *Beloved Beasts: Fighting for Life in an Age of Extinction* (New York: Norton, 2021).
2. Joseph Saragusty et al., "Rewinding the Process of Mammalian Extinction," *Zoo Biology* 35 (2016): 280–92.
3. For the purposes of this chapter, I am setting aside practical, scientific, moral, and ethical questions concerning whether this and similar projects should be undertaken at all. Since many of these projects are already underway, I am interested in their current and potential contributions to a wilder zoo and a wilder world.

4. Estimates of global failure rates for reintroductions range from 33 percent to 89 percent. See Jaume Badia-Boher et al., "Raptor Reintroductions: Cost-Effective Alternatives to Captive Breeding," *Animal Conservation* 25 (2022): 170–81.
5. Oliver A. Ryder et al., "Exploring the Limits of Saving a Subspecies: The Ethics and Social Dynamics of Restoring Northern White Rhinos (*Ceratotherium simum cottoni*)," *Conservation Science and Practice* 2 (2020): e241.
6. Onnie Byers et al., "The 'One Plan Approach': The Philosophy and Implementation of CBSG's Approach to Integrated Species Conservation Planning," *WAZA Magazine* 14 (2013): 2–5.
7. Ben A. Minteer, Jane Maienschein, and James P. Collins, "Zoo and Aquarium Conservation: Past, Present, Future," in *The Ark and Beyond: The Evolution of Zoo and Aquarium Conservation*, ed. B. A. Minteer, J. Maienschein, and J. P. Collins (Chicago: University of Chicago Press, 2018), 1–12.
8. See chapter 4, by Susan Clayton, in this volume.

Postscript

ON WILDNESS AND RESPONSIBILITY

BEN A. MINTEER AND HARRY W. GREENE

So many intriguing arguments, examples, and themes have emerged in the preceding pages that it would be a fool's errand to try to summarize all of them here. But to close this conversation, we'd like to briefly highlight a few lines of thought that will continue to shape our thinking about the possibilities and challenges of the wild in the zoo and especially about efforts to realize a "wilder zoo."

Clearly, most if not all of our authors believe the time has long since come to reassess the meaning of the wild, our relationships to and within it, and how zoos might become more wild-centric as part of this shifting understanding. In fact, one of the strongest conclusions emerging from this project is a need to bury the tradition of seeing zoos and the wild as an ecological and cultural binary operating at cross purposes. As an alternative, many of our contributors have explicitly emphasized notions of relationality—among humans, other animals, and the wild more generally—and the implications of this perspectival shift on how we conceive of and engage the zoo environment and its animal inhabitants.

A number of the discussions here have also shown how zoos could play a more significant role in wildlife and ecological conservation in situ, including by focusing their collections more deliberately on threatened species, facilitating the adaptation of animals under their care to a rapidly changing environment, and otherwise working diligently to create a future for imperiled species far beyond their walls. In these scenarios, the wildness of zoo animals becomes (perhaps ironically) redefined not as autonomy from human control but by the hands-on effort to help wildlife and the wild survive in an increasingly human shaped and ecologically turbulent age.

The image of zoos as a portal to the wild, one conveyed across many of the chapters in this volume, has long been part of the desired zoo experience, both for zoogoers and a zoo community desirous of scrubbing from public memory the older, unsavory, and "unnatural" exhibit aesthetic defined by cages and concrete. The "wilder," more naturalistic, and less patently artificial the zoo environment appears to visitors, the more progressive and ethically attuned its environmental philosophy also appears to be. Yet one particular observation, perhaps as obvious as it is provocative, has captured our attention as we come to the end of this project. The idea was in the air at our workshop at the Desert Museum in the fall of 2021 and has since burrowed its way into this book. It acknowledges that, despite the foregoing arguments for a naturalized, conservation-invested, and wilder zoological milieu, zoos are more for the people that visit them than for the animals they hold.

This conclusion underscores a tension we believe isn't fully resolved by efforts to relativize our notions of wildness or to promote a more relational vision of human culture and wild nature. And thinking of even a wilder zoo as ultimately more for people presents both an operational and a philosophical challenge for zoos and the nature and experience of the wild on display. Consider the following episodes:

- A father and his toddler daughter entered the elephant exhibit at the San Diego Zoo in search of an arresting selfie photo.

The elephant rushed them, causing the man to drop the child in his effort to escape. They got out unharmed, though it easily could have gone very differently.[1]

- A mother at the Cleveland Metroparks Zoo dangled her two-year-old son over the cheetah exhibit, and the boy fell into the enclosure, injuring his leg in the process. Luckily, the cheetahs seemed to ignore him, and he escaped without additional injury.[2]
- In Texas, a woman entered the monkey habitat at the El Paso Zoo and offered the animals Hot Cheetos. The encounter is posted on Instagram, which led swiftly to her firing by a publicly embarrassed employer.[3]
- In an especially dramatic incident, a woman breached the Bronx Zoo lion exhibit, getting within feet of one of the animals, waving her arms and appearing to have taunted it. "I wasn't fearing of the lion because the lion loved me," she later explained, describing the episode as a "spiritual" experience.[4]

Not surprisingly, selfies weren't the only arresting memories these trespassers took from their zoo visits. But what's interesting about these incidents, setting aside what we may infer about these individuals' states of mind (which ranges from reckless to what we might describe as disordered), is that they also manage to convey something deeper and maybe even paradoxical about zoos and their relationship to the wild. The desire to connect to the animal in the enclosure, to get a closer and better look, a meaningful if only fleeting encounter with that elephant, that monkey, that lion: it's something zoos have long encouraged, albeit in more ethically responsible, managed, and law-abiding ways. These intrusions, however, reveal a contradiction at the heart of this craving and in the project of making zoos wilder as part of an enhanced visitor experience. On the one hand, zoo animals are not entirely or truly "wild" compared to their counterparts in their natural habitats, however managed those habitats may be. But neither are they "tame" or domesticated.

On some level, these individuals and others like them who act on the inclination to illicitly enter animal exhibits must feel this, hence the thrill of stepping over the line and entering the animals' environment.[5] A memorable interaction with and digital record of the experience with an exotic animal is no doubt much of the draw, but so is the pull of the wild, however reduced, qualified, and curated it may be in the zoo context. Perhaps it is even a sense of being back *in* the wild, a yearning for being reciprocally part of wild nature, an ethic that would cede at least some control and will to the animals themselves.

We began this project with a long list of questions, and our authors have gamely answered most of them. But here at the end we find ourselves asking still more questions. Will visitors really develop more meaningful and ethically responsible connections to animals, wildlife, and larger ecological processes in a wilder zoo? If so, will these connections actually lead them to care more deeply about the fate of the wild beyond zoo gates, be it for lions in the Serengeti or peregrines in the midst of U.S. cities? Might visitor immersion in a wilder zoo even model a new environmental ethic of conscientious living alongside other species, perhaps a way of experiencing the wild that encourages respectful, bidirectional participation? That is, could zoos encourage an ethos defined not by entertainment preferences but by ecologically and culturally appropriate rules—of interspecies relationships, responsible animal engagements, and reciprocal taking and giving—in which visitors are no longer spectators but part of ecological processes? Or will a wilder zoo simply be a more amusing yet ultimately forgettable animal attraction in a world increasingly full of them? Only time will tell.

Yet we know from the conversations in these pages that zoos possess a still unrealized potential to connect their visitors to nature and to wildness in all shapes and sizes—from microbes to megafauna—in powerful and durable ways. We've also seen that these connections can be invigorated through a variety of creative means, from more engaged storytelling about wildlife and nature, to the accommodation and interpretation of a fuller array of natural animal behaviors in exhibits, to reforming the physical design of zoos to incorporate the conditions and experience of naturalness far more effectively. It's a hopeful inventory not only of the ecological and aesthetic possibilities still incipient within zoos today but of those within our collective environmental and cultural imagination as we work to realize a wilder zoo for tomorrow.

NOTES

1. "Man's Selfie Attempt with Daughter in San Diego Zoo Elephant Enclosure Nearly Gets Them Killed," CBSN Sacramento, March 22, 2021, https://gooddaysacramento.cbslocal.com/2021/03/22/elephant-enclosure-san-diego-zoo-selfie/.
2. Corinne Cathcart, "Boy Who Fell Into Cheetah Exhibit Was Dangled Over Railing, Zoo Says," *ABC News*, April 13, 2015, https://abcnews.go.com/US/boy-fell-cheetah-exhibit-dangled-railing-zoo/story?id=30259586.
3. Brittany Shammas, "Video Shows Woman Inside Zoo Exhibit, Where She Tried to Feed Cheetos to Monkeys," *Washington Post*, May 25, 2021, https://www.washingtonpost.com/nation/2021/05/25/el-paso-zoo-spider-monkeys-cheetos/.
4. Elisha Fieldstadt, "Woman Wanted for Climbing Into Bronx Zoo Lion Exhibit: 'I Am the Lion Now,'" *NBC News*, November 1, 2019, https://www.nbcnews.com/news/animal-news/woman-wanted-climbing-bronx-zoo-lion-exhibit-i-am-lion-n1075101.
5. Indeed, the impression of wildness is no small part of the allure of the zoo experience. As Irus Braverman has observed, "Without wildness, there would be nothing exciting about captivity; the zoo would be as domestic as a barnyard." Irus Braverman, *Zooland: The Institution of Captivity* (Palo Alto, CA: Stanford University Press, 2013), 60.

ACKNOWLEDGMENTS

First and foremost, we'd like to thank our authors, each of whom engaged our questions about zoos and the wild with great energy, collegiality, and insight. A special thank you goes out to Craig Ivanyi and Debbie Colodner for hosting our group in a very memorable and productive workshop at the Arizona-Sonora Desert Museum in November 2021. We're also grateful to Jessica Ranney and Elena Islas at Arizona State University's Center for Biology and Society and Maria Castano at the Desert Museum for providing key logistical support for the workshop.

We were thrilled to have Miranda Martin, our wonderful editor at Columbia University Press, at the Desert Museum gathering; we appreciate her strong support of this project through its many phases. We're also indebted to three excellent PhD candidates in Biology and Society at ASU, Daniel Bisgrove, Olivia Davis, and Cassandra Lyon, who played key roles at the workshop. Finally, essential financial support for the project was made possible by Ben Minteer's Arizona Zoological Society Endowed Chair position at ASU.

CONTRIBUTORS

IRUS BRAVERMAN is professor of law and adjunct professor of geography at the State University of New York at Buffalo. Her books and edited volumes include *Zooland: The Institution of Nature* (2012); *Wild Life: The Institution of Nature* (2015); *Animals, Biopolitics, Law: Lively Legalities* (2016); *Gene Editing, Law, and the Environment: Life Beyond the Human* (2017); *Coral Whisperers: Scientists on the Brink* (2018); *Blue Legalities: The Laws and Life of the Sea* (2020); and *Zoo Veterinarians: Governing Care on a Diseased Planet* (2021). Braverman's forthcoming book is titled *Settling Nature: The Conservation Regime in Palestine-Israel*. She is involved in multiple transdisciplinary conversations and forums with conservation biologists, veterinarians, and geneticists.

SUSAN CLAYTON is Whitmore-Williams Professor of Psychology at the College of Wooster in Ohio. Her research, which has been conducted at several zoos around the world, focuses on the human relationship with nature, how it is socially constructed, and how it can be used to promote environmental conservation. She has also written extensively about the implications of climate change for human well-being. Her most recent book is *Psychology and Climate Change: Human Perceptions, Impacts, and Responses* (2018).

DEBRA COLODNER is Director of Conservation Education and Science at the Arizona-Sonora Desert Museum, where she oversees interpretive and educational programs, conservation science and outreach, as well as the natural history collections. She applies her passion for informal science learning and community-based conservation to the development of programs and partnerships in support of Sonoran Desert ecosystems and the people that depend on them.

ALISON HAWTHORNE DEMING is Regents Professor Emeritus at the University of Arizona. She is the former director of the UA Poetry Center, and her work has been awarded fellowships from the Guggenheim Foundation, Walt Whitman Award from the American Academy of Poets, Stegner Fellowship, and the National Endowment for the Arts, among other honors. Her most recent books are *A Woven World* (2021); the poetry collection *Stairway to Heaven* (2016); the essay collection *Zoologies: On Animals and the Human Spirit* (2014); and *Death Valley: Painted Light* (2016), a collaboration with the photographer Stephen Strom. Her work is widely published and anthologized, including in *The Norton Book of Nature Writing* and *Best American Science and Nature Writing*.

HARRY W. GREENE taught for two decades at the University of California, Berkeley, and is now professor emeritus of ecology and evolutionary biology and Stephen H. Weise Presidential Fellow at Cornell University and adjunct professor of integrative biology at the University of Texas at Austin. His honors include the top teaching awards at Berkeley and Cornell and election to the American Academy of Arts and Sciences. His *Snakes: The Evolution of Mystery in Nature* (1997), won a PEN Literary Award and made the *New York Times*' "100 Most Notable Books" list, and the more recent *Tracks and Shadows: Field Biology as Art* (2013) was highly favorably reviewed in *Publishers Weekly*, *Booklist*, *Science*, *Nature*, *Current Biology*, *The Sciences*, *Natural History*, and *Times Higher Education*.

CRAIG IVANYI is the executive director of the Arizona-Sonora Desert Museum, a position he's held since 2010. Much of his career has focused on the natural history and conservation of lower vertebrates of the Sonoran Desert region, with particular emphasis on venomous reptiles.

KELSEY DAYLE JOHN (Navajo) is an assistant professor at the University of Arizona with a joint appointment in American Indian studies and gender and women's studies. She studies equine/human relationships in tribal communities with a focus on the social, cultural, and historical narratives of equine and human co-constructed living. Alongside her work in Indigenous animal studies, Kelsey's research interests also include Indigenous feminist studies, American Indian studies, and foundations of education. She serves on the board of trustees for the Arizona-Sonora Desert Museum. She has published in the *American Indian Culture and Research Journal*, *Humananimilia*, *Edge Effects*, and several edited volumes.

JONATHAN B. LOSOS is an evolutionary biologist known for his research on how lizards rapidly evolve to adapt to changing environments. He is the William H. Danforth Distinguished University Professor at Washington University and director of the Living Earth Collaborative, a biodiversity partnership between the university, the Saint Louis Zoo, and the Missouri Botanical Garden. Losos has written more than 240 papers and three books and is an author of a leading college biology textbook. Losos has been elected a member of the National Academy of Sciences and a fellow of the American Academy of Arts & Sciences and is the recipient of a Guggenheim Fellowship and many other awards.

CASSANDRA LYON is a PhD candidate in the Biology and Society program in the School of Life Sciences at Arizona State University. She is broadly interested in human-animal interactions studies, with a particular interest in zoos. Her dissertation research explores the presence of connection messaging, such as anthropomorphism, and conservation messaging in zoo exhibits. She also studies how human actions, perceptions, and assumptions influence the management of domestic and exotic animals in human care and in natural ecosystems and animals living in the wild.

CURT MEINE is a conservation biologist, environmental historian, and writer based in Sauk County, Wisconsin. He serves as senior fellow with the Aldo Leopold Foundation and Center for Humans and Nature; as research associate with the International Crane Foundation; and as adjunct associate professor at the University of Wisconsin–Madison. Meine has authored and edited several books, including the award-winning biography *Aldo Leopold: His Life and Work* and *The Driftless Reader*. He served as on-screen guide in the Emmy Award–winning documentary film *Green Fire: Aldo Leopold and a Land Ethic for Our Time* (2011). In his home landscape, he is a founding member of the Sauk Prairie Conservation Alliance.

NATASCHA MEUSER is a professor at the Anhalt University of Applied Sciences (Germany). Meuser is the author of numerous publications in the field of design methodology and building history, as well as architecture and zoology. In 2020 she established the Institute of Zoo Architecture (ZooArc) at the Anhalt University of Applied Sciences. She is the author of *Zoo Buildings: Construction and Design Manual*, the first such book to explore zoos as a mode of architecture.

JOSEPH R. MENDELSON III is director of research at Zoo Atlanta and adjunct professor at the Georgia Institute of Technology. His research is field based and lab based, with much of the lab work conducted at Zoo Atlanta using its living collections. As a herpetologist, he has published widely on topics such as the behavior, anatomy, evolution, biogeography, taxonomy, conservation, and natural history of amphibians and reptiles. He also has published papers addressing various aspects of the concept of zoos as research organizations. His work has appeared in *Science*, *Proceedings of the National Academy of Science*, *Herpetological Review*, *Journal of Experimental Biology*, and several edited volumes.

BEN A. MINTEER is professor of environmental ethics and conservation and the Arizona Zoological Society Endowed Chair in the School of Life Sciences at Arizona State University. Minteer's work has appeared in *Science*, *Nature*, *BioScience*, *Slate*, and *Earth Island Journal*, among other outlets. He has published many books, including *The Fall of the Wild: Extinction, De-Extinction, and the Ethics of Conservation* (Columbia, 2018) and *The Ark and Beyond: The Evolution of Zoo and Aquarium Conservation* (2018; *Choice* Outstanding Academic Title). His most recent book (coauthored with Mark Klett and Stephen J. Pyne) is *Wild Visions: Wilderness as Image and Idea* (2022).

HOLLY G. MOLINARO is an animal welfare scientist, with research experience in zoos on species ranging from baboons to crocodiles. She is currently a PhD student in psychology at Arizona State University, studying dogs' positive emotional states and how humans understand happiness in dogs.

GARY PAUL NABHAN is an ethnobiologist, contemplative ecologist, conservation biologist, and biocultural geographer trained at the University of

Arizona and Prescott College. He is the author or editor of more than thirty books translated into six languages, a number of which have won awards. In addition to his research, teaching, and community service on sustainable food systems, Nabhan has worked in zoos and botanical gardens and in habitat restoration in the desert and seas. A MacArthur Fellow and prolific author, his most recent books are *Agave Spirits* (2023) and *The Nature of Desert Nature* (2020).

MICHELLE NIJHUIS, a project editor for the *Atlantic* and a longtime contributing editor for *High Country News*, writes about science and the environment for publications including *National Geographic* and the *New York Times Magazine*. After fifteen years off the electrical grid in rural Colorado, she and her family now live in southwestern Washington. Her book *Beloved Beasts: Fighting for Life in an Age of Extinction* was published in 2021.

NIGEL ROTHFELS is a historian of animals and culture. He is the author of a history of naturalistic displays in zoos, *Savages and Beasts: The Birth of the Modern Zoo* (2002); the editor of a multidisciplinary collection of essays in animal studies, *Representing Animals* (2002); a coauthor of a study of elephants and keepers at a contemporary American zoo, *Elephant House* (2015); and the author most recently of a book on the history of ideas about elephants, *Elephant Trails* (2021). Rothfels is also the general editor of a book series in animal studies with Penn State University Press, Animalibus: Of Animals and Cultures, with nineteen volumes in print.

REVA MARIAH SHIELDCHIEF is a PhD candidate in American Indian studies at the University of Arizona. She was first an instructor for writing and later became an American Indian studies and adjunct Tohono O'odham studies instructor at Tohono O'odham Community College. She also served as the American Indian Studies Department chair at Pawnee Nation College. Her poetry and short stories have appeared in anthologies such as *Sister Nations* and journals such as *Red Ink: A Native American Student Publication*, as well as the *New York Review* and *Cimarron Review*.

AMANDA STRONZA is an environmental anthropologist and professional photographer with thirty years of research and conservation work in the

Amazon, the Okavango Delta, and other parts of the tropics. She is a professor at Texas A&M University in the Department of Ecology and Conservation Biology. Stronza cofounded Ecoexist, a nonprofit organization in Botswana aimed at fostering coexistence between people and elephants. She is a recipient of the national Praxis Award in Anthropology for her work in translating anthropological knowledge into concrete action to support community conservation and development in Africa and Latin America.

CLIVE D. L. WYNNE is the founding director of the Canine Science Collaboratory at Arizona State University. Previously, he was founding director of the Canine Cognition and Behavior Laboratory at the University of Florida, the first lab of its kind in the United States. The author of *Dog Is Love* (2019) as well as several previous academic books and more than one hundred peer-reviewed scientific journal articles that count among the most highly cited studies on dog psychology, he has also published pieces in *Psychology Today*, *New Scientist*, and the *New York Times* and has appeared in several television documentaries about dog science on National Geographic Explorer, PBS, and the BBC.

INDEX

Africa: rhino conservation in, 232, 234; wildness, human influence and, 15–16. *See also* South Africa
agency, wildness, and, 37, 80, 88–89
Albatross, 53
Amazon, 169; human-nature interaction in, 19–20; manatee tank in Columbian zoo of, 218–19
ambassador animals, 141, 199
American black bears, feeding time and, 161–62
American Indians. *See wild*, American Indian understanding of
Anhalt University of Applied Sciences, in Dessau, 194
animals: ambassador, 141, 199; anthropomorphized, 98; art using taxidermied, 53; behavior as wildness, 116; cathemeral, 162; cave paintings of, 51–52; close-up experiences with, 117–18, 121; crepuscular, 162; demonstrations of, 117–18; desire for connection with wild, 66–67; emotional response to, 73; empathy with, 199; enclosure size, 118–19, 120, 121; exhibits of nocturnal, 156–60, *157*, *159*; living beings with rights view of, 193; management required for zoo *wild*, 119, *119*; *New Yorker* cartoons of, 56; as persons, 80–81; prioritizing exotic, 200–201; rescued, 110; restraining of, 162; "risky" contact with, 95; selfies with, 242–43; social imprinting in, 143–44; zoo design based on, 164–66; zoos creating empathy with, 199. *See also* coexistence, human-wildlife
animals, captive: antipredator behavior and, 128–29, 132–35; birth management of, 162–63; connection with, 63, 64–65;

animals, captive (*continued*)
debates over zoos and, 3, 214–15; deformed specimens of Peter the Great, 96; empathy with, 65; evolutionary pace context for helping, 125–26, 136n1; First World dogs as, 147–50, *149*; first zoos containing large, 44n7; food as means of controlling, 161–62; individual stories of, 59; loss of the wild in, 95; negative duties to, 7; positive duties to, 6–7; replacement and names of dead, 163; sequestering of, by type, 162; visibility, temporality and, 165–66; visitor trespassing incidents with, 242–44, 245n5
animal welfare: architecture focus on, 189–90; five freedoms principle of, 148–49; visitor vs., 189–90, 192, 193
Anthropocene, 11; "wild" as diminished in, 49
antihunting agendas, 18
antipredator behavior: boodies and bilbies experiment on, 132–35; loss of, 128–29
aquariums, 172; annual visitor numbers, 186–87; as climate protection stages, 184–88; conservation budget of zoos, gardens and, 102–3; construction boom in, 187; development of zoos and, 95–98; development plans for new, *187*, *190*, *191*, *192*; importance of zoos and, 111
architecture, zoo and aquarium: animal-human coexistence context for, 183–84; animal welfare focus of, 189–90; building typology and, 194; educational tasks of, 183–84, 193–94; literature absence, 188; specialists, 183; structural infrastructure changes for, 189–90; wildness, academicization and, 193–94
Arid Recovery, 132–35
Arizona-Sonora Desert Museum (ASDM), 2, 43n4, *113*; accredited status of, 110; annual visitors, 109, *109*; characteristics, 25; cofounder vision for, 100–101; connection to nature aspect of, 64; establishment of, 25; evolution and expansion of, 109; exhibits, 26; former director on conservation budgets, 102–3; habitat-based exhibits of, 100; idealized *wild* of, 122; institutional distinctions erased by, 179; introduction to, 107–12; lizards in, 118, 124n6; location, *108*; Luke "hyperreality" critique of, 101, 109; main attraction, 109; making wilder, 118–20, 122; mission of, 108–9; outdoor exhibits of, 95; Raptor Free Flight program, 119, *119*; rescued animals of, 110; species recovery focus of, 100; study

participation of, 111; survey, 111–15, 124n6; 2021 gathering of scholars, 93; Wilder Kingdom workshop, 58–59, 107–8, 111; zoo design, 26; zooness question and responses, 112, *114*, 114–15, 123n4. *See also* survey, ASDM
art: cave, 51–52; empathy invoked by, 59–60; environmental, 55; Nevada Museum of Art, 53, *54*; origin of, 51–52; science and, 50, 51–52, 58; wilder zoos project, 58–59; in zoos, 52
artificial insemination, 229–30
artificial selection, 131, 132, 145; Darwin theory of unconscious, 142; dogs and, 142
ASDM. *See* Arizona-Sonora Desert Museum
Asian elephants, in zoo, 57
Association of Zoos and Aquariums (AZA), 34, 63, 201, 215–16; accreditation requirements and, 202
Australia, 141; marsupial cats from, 128; predator removal in, 131; quolls moved to, 128; research on predators and kangaroos, 132–35
autoclave, 178
AZA. *See* Association of Zoos and Aquariums

bandicoots, hare-wallabies and, 131
Barongi, Rick, 199
Bear Mountain Trailside Museums, 101
bears, 10, *10*; London Zoo bear pit, 161–62, 164; respecting agency of bear-persons, 80
behavioral ecology, 15
Berlin Zoo, elephants named Roland in, 163
biophilia, 67, 99
births, of zoo animals, 162–63
bison, 8, 231
boodies and bilbies, Australian experiment with predators and, 132–35
botanical gardens: ASDM, *113*, 114; development of zoos and, 95–98
breeding, endangered species, 130–31. *See also* captive breeding
Bronx Zoo, 8, 9, 164; lighting experiments, 157–58; nocturnal exhibit, 157–58; small enclosures of past, 218; "spiritual" experience with lion in, 243
Brookfield Zoo, 38, 101
Brower, David, 1, 19, 205, 206
Buikx, Jasper, 172, 176, 178
butterfly exhibits, 163

cactus, Thornber fishook, 102
cages and enclosures: enclosure of the wild, 93–95, 101; examples of too small, 218–19; human-nature divide and, 217; macaques in portable, 219; reforms to, 219–20; size of, 118–19, 120, 121; Woodland Park Zoo, *10*; Zurich elephant, 190

California condors, 199, 209, 231–32; captive breeding of, 204–5; human intervention and, 205–6
Canadian Maritimes, 47–48
captive breeding: artificial insemination and, 229–30; California condor program of, 204–5; "dignity" narrative against, 198, 205, 207–10; of microbes, 176–77; reintroduction and, 231, 233; species benefiting from, 214; treefrog program, 207, 208, *208*; uncertainty and distorted views of, 235–36
captive populations: adaptation to zoo environment, 127; artificial selection and, 131; natural selection and, 125–30, 132–36; reintroduction to natural environment, 127–28, 136; traditional approach to managing, 129. *See also* animals
captive populations, new evolutionary approach to: hypothetical example of lizards and, 130–31
Caretta Project, 203–4
Carr, Bill, 100–101
cathemeral animals, 162
cats: Australian foxes and, 131; in boodies/bilbies experiment with predator, 132–35; coyotes preying on, 60–61; songbirds killed by domestic, 61
cave art, 51–52
Central Park, 8, 28n19
Channel Island, foxes of, 19
Chauvet cave, 51
cheetah enclosures, dogs in, 139, *140*
children, 66, 95; childcare issue and, 188–89; direct interaction with nature, 99; false sense of *wild* in, 98–99; Pokémon and, 99; studies on, 99
chimera, 53, *54*, 55, 60
China: aquarium popularity in, 186–87; dogs and, 140; segregation of walled and wild in, 36–37
Christianity, of EuroAmerican settlers, 82–83, 90n1
Cincinnati Zoo, last passenger pigeon death in, 101
Civil War, 84, 97
Clark, Kate, 53–55, *54*
Cleveland Metroparks Zoo, 243
Climate Accord congress, in Glasgow, 93
climate change: aquariums and, 184–88; Greenland glaciers example of, 184–85
Club of Rome, 185
coexistence, human-wildlife, 238–39; architecture and, 183–84; cranes example of, 41; rural people-elephant, 222–23, 225; strategies, 213
Cologne Zoo, 189, *190*, 191, *191*
colonialism, 237; Christianity and, 82–83, 90n1; "dignity" and, 205–6; of first zoos, 36, 37; fortress model of, 37; *wild*, noble

savage trope and, 81–85; *wilder* kingdom and, 88–89
Colossal, 235
compassion, 20, 122
condors. *See* California condors
Connect, 63
connection: with captive animals, 63, 64–65; conservation support through, 64, 70; desire for connection with wild, 66–67; with elephants, 217, *217*; to nature, 64–70; new stories for zoos and parks, 223–25; zoo design goal of, 68–71; zoos and park separation and, 216–19
conservation: coexistence and, 238–39; coexistence strategies for, 213; crane, 31; dualisms in, 37; elephant, 220; elephants and keepers in Thailand centers, 69, *69*; empathy with zoo animals as help to, 199; ex situ, 102–3, 235; ex situ-in situ integration, 236–39; funding, 102–3, 235; genetics, 229–30, 231–32; hands-off, 204–6; human-nature divide in, 223–24; ideology clash with, 204–6; in situ, 199, 202; *inter-situ*, 34, 38, 41, 42; of local species, 202–3; misperceptions about people and, 238–39; movement goals, 126; northern white rhino project and, 229–30; overseas, 201; predator-removal approach to, 131–32; regional, 202–4, 208–9; rhino, 232, 234; sense of connection leading to support of, 64, 70; from shared identity with individual animals, 122; traditional, 37; trans-situ, 95, 100; WAZA focus on, 193; zoo involvement in, 231; zoo messaging and public perception of, 232–36. *See also* captive breeding; *specific species*
conservation biology, field of, 126–27, 171
conservation history: environmental ethics and, 5–6; zoos as absent in, 7–8, 28n16
Conway, William, 38, 198–200, 201; applying vision of, 204, 209–10
coral reef, 236
Cousteau, Jacques-Yves, 185
"Crane City," 32
cranes: coexistence of humans and, 41; dialogue between wild and captive, 31, 32, 33, 34, 39; family name, 31; Ho-Chunk name for, 42; most abundant species of, 35, 40; specialized, 40; varied voices of, 32
crepuscular animals, 162
crises, financial and ethical, 93–95
Cronon, William, 117
culture: building, 193–94; zoo dichotomy of nature and, 87

Darwin, Charles, 125–26, 136n1, 142
decolonization, of museums, 86–87
"de-extinction," 235

deformed specimens, of Peter the Great, 96
Deloria, Vine, Jr., 83, 89
Deming, Alison Hawthorne, 50–51
"Desert Reservation" (Lopez), 93–94
Different Nature, A (Hancocks), 2
dignity: captive breeding and, 198, 205, 207–10; as human construct, 205–6
dingos, 141
Disney, Walt, 98
dogs: as cheetah companions, 139, *140*; domestication of, 142–44; fed to snakes, 141; First World pet, 147–50, *149*; five freedoms and, 148–49; in human dwellings, 144–45; misunderstandings about, 142–47; as natural selection outcome, 142–44, 147; percentage living outside human homes, 145; rescued, 148; tame v. domesticated, 146; trapped in homes, 147–50, *149*; wolves becoming, 142–43; zoos and, 139–41, *140*, 147
domestication: of dogs, 142–44; evolutionary change and, 127–28; social imprinting and, 143–44, 146–47; songbirds killed by domestic cats, 61; "tame" distinguished from, 146; *wild* as not opposite of, 145–47
During the Rut (Rungius), 52
Ecker, Liz, 102
eco-in-centricity, 22
ecological interactions: ASDM outdoor exhibits allowing, 95; death of species from lack of, 101; isolation and lack of, 97–98
ecology, behavioral, 15
ecotourism, 121
Egypt, 49
Elephant Crisis Fund, 222
elephants: Berlin Zoo, 163; coexistence of rural people, 222–23, 225; conservation, 220; country with largest population of, 222; experience of connecting with, 217, *217*; human conflict with, 221–22; hybrid embryo of, 235; ivory trade killing of, 222; life span of, 215; Little Rock Zoo Asian, 57; movement corridors of, 224–25; selfie attempt with, 242–43; space requirements, 215–16; stories and, 220–21, 222–23; in Thailand conservation centers, 69, *69*; wilder, wider angle on, 221–23; wilder zoos and parks implications for, 220–21; Zurich enclosure for, 190
El Paso Zoo, 243
empathy: art and, 59–60; conservation and, 199; environmental protection and, 65
enclosures. *See* cages and enclosures

endangered species: assistance opposition argument, 198; breeding, 130–31; Conway on zoos centered on, 198–200; cranes as most studied, 33; "dignity" and, 198, 205–10; earlier focus on genetic variation within, 126–27; location of most, 201; natural selection experiment and, 132–35; red list of, 32; wilder zoos and, 242. *See also* extinction
environmental art, 55
environmental ethics: fields of conservation history and, 5–6; "good-bad" zoo question and, 12–13; zoo issues absence in, 4, 7
Ethnobiology for the Future (Nabhan), 99
Eurasian crane, 40
EuroAmerican settlers, Christian morals of, 82–83, 90n1
evolutionary change: artificial selection and, 131; Darwin's mistake about, 125–26, 136n1; pace of, 125–26, 128–29, 134, 143; traditional approach to captive populations and, 129
extinction: "de-extinction" trend, 235; with dignity, 198, 205, 207–10; of experience, 98

feeding time, 160–63, *161*
Foucault, Michel, 170
foxes, of Channel Island, 19
Frankel, Otto, 126
French kissing, 176
Frozen Zoo, 232

galagoes, 157
"Gardens and Menagerie of the Zoological Society of London," 35–36
General Allotment Act, 84
genetic drift, 127
genetics, conservation, 229–30, 231–32; hybrid embryo of elephant and, 235
genetic variation, conservation focus on, 126–27
Germany, Magdeburg Zoo in, 191, *191*
glaciers, Greenland, 184–85
God of Small Things, The (Roy), 179–80
Gondwanaland, 190
gorillas, 160, 207
Grand Manan Island, 47–48
Grazian, David, 11
Great Barrier Reef, 236
Great Britain, zoo development in, 95–97
Great Plains, 19
Great White Hunter, 103
grebe, Atitlan flightless, 207
greenhouse gases, 186
Greenland glaciers, 184–85
grief, wonder and, 59
Guatemala, grebe (*poc*) of, 207, 209

Hagenbeck, Carl, 11, 140–41, 163
Hancocks, David, 2, 13–14
hare-wallabies, bandicoots and, 131

Hawaii Rare Plant Facility, 101
HEC. *See* human-elephant conflict
Hediger, Heini, 164–65
Hirst, Damien, 53
Ho-Chunk Nation, 41–42
hominins, 17, 20
Hornaday, William T., 8, 231
horses, 83; in cave drawings, 51; last wild subspecies of, 51; southern Great plains Indians and, 19; wild mustangs, 78, 79
human-elephant conflict (HEC), 221–22
human intervention: California condor success and, 205–6; climate change and, 185–86; "dignity" belief in opposition to, 198, 205, 207–10; domestication mistaken association with, 142–44; required, 73
humans: benchmark for "no human impact," 16; dogs living with, 144–45; elephants and, 69, *69*; elephants coexistence with rural people, 222–23, 225; history of nature-human involvement, 19–20; irony of close animal-human interactions/*wild*, 119, 121; microbiome and, 169; minimization of nature involvement of, 19–20; as part of ecosystem, 72–73; percentage of dogs living outside of human homes, 145; species more like, 215; *wild* as absence of, 16, 116, 117; "the wild" definitions and acknowledgment of, 72–73; wilderness values threatened by, 23; wildness, agency and, 37; wildness dilemma in light of, 15–16; wildness participation of, 18, 22. *See also* coexistence, human-wildlife
hunting, 16, 17–18, 35
Huston, Karla, 58
Hyson, Jeffrey, 9

IAN/IAC. *See* International Aquarium Network
ICF. *See* International Crane Foundation
immersive exhibits, 9, 11, 163–64
Industrial Revolution, 95–98, 101; children lack of nature experiences from, 99; time concepts based on, 155–56, 161
Inert Wolf, 53
institutions: ASDM and, 179; conservation budgets of ex situ, 102–3; crises, 93–95; dilemma of, 188; Great Britain's nature-derived, 95–97; institutionalization start, 95–98
instrumentalist values, intrinsic v., 18, 21
International Aquarium Network (IAN/IAC), 187
International Crane Foundation (ICF), 31–32, 34–36; Conway praise of, 199; crane pens and murals, 40–41; education programs, 40; founding of, 33;

Ho-Chunk Nation and, 41–42; recent renovation, 42
International Rhino Foundation, 237–38
International Union for Conservation of Nature (IUCN): Elephant Specialist Group, 222; One Plan, 233; Red List of Threatened Species, 32
inter-situ sites, 38, 41, 42; ICF as, 34
intrinsic values, 20; instrumentalist v., 18, 21
isolation, 99–100, 176–77; "saving" species in, 103
IUCN. *See* International Union for Conservation of Nature

Jordan, Chris, 53

Kabul, Afghanistan, dogs rescued from, 148
kangaroos, Australian boodies and bilbies, 132–35
Kenya, 15, 23; Northern Rangelands Trust in, 237; northern white rhino in, 236
Kew Gardens, 96–97
kiss-o-meter, Micropia's, 176
Kort, Remco, 172, 173–75, 178
Kruger National Park, in South Africa, rhinos in, 23–24

lions, 160, 243
Little Rock Zoo, Asian elephants encounter in, 57
lizards, optimal temperature, evolution and, 130–31, 132
London Zoo (London Zoological Gardens), 35, 36, 96–97; feeding schedule, 160–61, *161*; zoo purpose and, 49–50
Lopez, Barry, 93–94
Louis XIV, menagerie at Versailles, 164
Lucy, fossil of, 17
Luger, Cannupa Hanska, 55
Luke, Timothy, 101, 109
Lyell, Charles, 125

macaques, in portable cages, 219
Madagascar Rainforest exhibit, 164
Magdeburg, 191, *191*
Man and Nature (Perkins), 97
manatee, small tank for, 218–19
Marsh, George Perkins, 97
Marshall, Alfred, 97
marsupial cats. *See* quolls
media: digital, 98; on elephants, 222; nature images in, 99; zoo messaging and, 232–36
menageries, 49–50; eighteenth-century, 140; Louis XIV, 164
microbiome, in humans, 169
Micropia-ARTIS, *175*; human microbiome and, 169; life span of, 174; making microbes die, 177–78; making microbes live, 173–77; microbes evolutionary success and, 171; microscope technology and, 173; number of Earth species of, 170–71; number of microbe species exhibited by, 172

Midway Atoll islands, 53
Milwaukee County Zoo, *157*
Milwaukee Public Library, 57–58, *58*
Milwaukee Public Museum, 57–58, *58*
minirevolutions, zoo design in, 9, 11, 163–64
monkeys, portable cages confining macaque, 219
Montezuma's zoo, 36–37
Muir, John, 20–21, 22
Museum of Anthropology and Ethnography, Peter the Great, 95–96
museums: decolonization of, 86–87; nature critique by art, 53, *54*; without walls, 103
museums, microbe. *See* Micropia-ARTIS
"myth of the pristine," 37

Nabhan, Gary Paul, 99
Namibia, 237
Nash, Roderick, 7
National Geographic, 102–3
Native Americans, 16
"natural," wild vs., 179
natural history: ASDM, 107; cave drawings as, 51–52; first book on, 96; pioneer of outdoor, 100–101; serpents in, 15; wildness as revealed by, 15–22
naturalism, wildness term in context of, 14–16
natural selection, 125; adaptive solutions based on, 127–28; boodies/bilbies/cats experiment in, 132–35; cats as agents of, 134–36; directional evolutionary change and, 126; dogs as outcome of, 142–44, 147; positive aspect of, 129–30; wolf-dog adaptations through, 142–44
nature: actual history of human involvement in, 19–20; awe-inspiring, 67; colonial ideas of "close to," 85–86; connection to, 64–70; contact with "risky," 101; as in continual flux, 129–30; decline in youth experience of, 98–99; digitization and, 191; direct interaction of children with, 99–100; ethos of people-less, 197; hyperreal images of, 98; minimizing human involvement with, 19; recent experiences in, 115–16; reverence for, 22; sanitized view of, 120; surrogates, 97, 101; urban places reclaiming by, 67; zoo dichotomy of culture and, 87
nature-derived institutions, development in Great Britain of, 95–97
Nature Fakers, 98
Nature Study Movement, 100–101
Navajo tribe, "wild" use by, 78–79
Nevada Museum of Art, 53, *54*
noble savage trope, 81–85
nocturnal exhibits, *157*, *159*; first, 156–58; reversed lighting in, 156–59
Northern Rangelands Trust, 237

northern white rhino project, 229–30; baby "Edward" and, 229, *231*, 233–34; Frozen Zoo collection of cells from, 232; messaging and public perception, 232–36
Norton, Bryan, 6, 67

oceans, 38, 207; Cousteau raising awareness about, 185, 187, *187*; greenhouse gases as impacting, 186; microbes and, 171; salinity decrease of Atlantic, 184–85
Old Testament, 82
Olmsted, Frederick Law, 8
One Plan, 233
On the Origin of Species (Darwin), 125–26, 142
ontogeny, "tame" linked with, 146
owls, great-horned, 119, *119*

Pagel, Theo, 189, 191, 193
Palmer, Clare, 6
pandemic, 42, 48, 188
paradise, 36
parks: first zoological, 97; irony of national, 86; managed wildness of, 197–98; movement, 8; wilder zoos and, 220–21. *See also* zoos and parks
Partulid snail, 169
passenger pigeons, 101
Peirce, Charles S., 5
People, 233
persons, animals as, 80–81
Peter the Great, 95–96
petri dish, of bacteria, *175*
phylogeny, domesticating and taming in, 146–47
Pleistocene, artists of, 24
poaching, 236–37
poc (grebe), 207, 209
poetry: "Desert Reservation," 93–94; installations, 56–58; path, 56, *58*; storytelling and, 56
Pokémon, 99
pragmatic preservation, 12–14; approach definition, 5; coinage of, 4–5; environmental ethics and, 4–7; "natural" and "wild" zoos and, 11; "pragmaticism" and, 5
predators: experiment with kangaroos and cats, 132–35; introduced, 131; removal of, 131–32
"Prelude" (Wordsworth), 60
preservation: destruction connection with, 85; ideology opposing zoos for, 204–6; policy goals of nature, 4; purists on, 206, 207–9. *See also* pragmatic preservation
preservationism, 18–19
Putnam, Walter, 85–86

quolls (marsupial cats), 128–29

Rabb, George, 38, 101, 200; treefrogs project and, 207, 208, *208*
radio tags, 205

rapid evolutionary change, 126, 128–29, 134, 143
Raptor Free Flight, 119, *119*
Rauschenberg, Robert, 53
"Red Light Room," 157–58
relationality (relationship): between and among species, 100–104; wildness, agency and, 88–89; zoo-wild, 1, 4–14, 33–34
Research Center of the Seas of Cortez, 187, *187*
revolution, wilder zoo, 163–66
rhinoceros, 206; black, 237; encounter with, 23–24; famous baby, 229, *231*, 233–34; Namibian success in saving, 237; northern white, 229, 230; poaching, 236–37
Roy, Arundhati, 179–80
Rungius, Carl, 52
Ryder, Oliver, 231, 232, 234, 237–38

Saguaro National Park, 109
Saint Louis Zoo, 209–10
salamanders, 209–10
salmon, in captivity, 129
sandhill cranes, 31, *35*, 39–40; near extinction and revival of, 34–35
San Diego Zoo Safari Park, 130; baby rhino in, 229, *231*, 233–34; communications staff, 233–34; dogs exhibited in, 151n9; rhino-conservation support by, 237; selfie attempt with elephant in, 242–43
Sarus cranes, 40
Schoenberger, Elisa, 86–87
Schweitzer, Albert, 20, 21
Science, 99
science, art and, 50, 51–52, 58
"Sea Change" (Huston), 58
Seattle, coyotes in, 60–61
sea turtles, 203–4, 207
selfies, with animals, 242–43
Serengeti, 21
serpents (snakes): annual deaths from bites, 23; arms race with, 17; constrictor, 16; dogs fed to, 141; natural history on, 15; persecution of, 21
Shepard, Paul, 94
She-Wolf, 55
Sierra Club, 20–21
size, wildness and, 169–70
"skidi" (wolf), 89n1
Skidi-Pawnee tribe, 79–80, 89n1
Smithsonian National Zoo, nocturnal room in 1980s, 158
snails, Partulid, 169
snakes. *See* serpents
social imprinting, 143–44, 146–47
soil microbes, 170–71
songbirds, killed by domestic cats, 61
Soulé, Michael E., 20, 126
South Africa: rhino conservation organization in, 237–38; rhino encounter in, 23–24
species: coexistence of humans with other, 213–14, 238–39, 244; death of surrogates for wild,

101; enclosure size, welfare of, 121; exclusion of nonendangered, 199; local, 202–3; more-like-human, 215; public belief in "saving" isolated, 103
Standing Rock, 55
Stegner, Wallace, 1
storytelling, poetry and, 56
survey, ASDM: conclusion from results, 120–23; contradictory result, 121; invitation and respondents, 111–12; making zoo *wilder*, 118–20; nature experiences, 115–16; responses, 112–15; "what do you like most" responses, *113*; on *wild* concept, 116–20, 124n6; on zooness of, 112, *114*, 114–15, 123n4

Taliban, dogs rescued from, 148
Texas, 16, 17–18, 243
Thailand, elephant conservation centers in, 69, *69*
Thoreau, Henry David, 22
Thornber fishook cactus, 102
Tibetan macaques, 102
Tibetan mastiff, 140
Tierpark, 163
time, zoo: feeding time, 160–63; industrialization context, 155–56, 161
trail, in Canadian Maritimes, 47–48
trains, zoos and, 155–56
treefrogs, Rabb fringe-limbed, 207, 208, *208*
Triennial Art + Environment Conference, 53
"Trouble with Wilderness, The" (Cronon), 117

unconscious selection, 142
United Kingdom, dog rescue controversy in, 148
United Nations Framework Convention on Climate Change, 184
United States: Industrial Revolution and institutionalized nature in, 97–98; wildness and natural habitat trend in, 194
U.S. Botanic Garden, 97

vanishing Indian, myth of, 84–85
Vermeij, Mark, 236
vespertine animals, 162
visibility: animal, 165; of living microbes, technology and, 176
visitors, zoo: child, 66, 95, 98–99, 188–89; desire for connection with wild animals, 66–67; focus on animals v., 189–90, 192, 193; instilling ecosystem membership of, 239; selfies and, 242–43; surveys of, 64, 65; trespassing and dangerous incidents, 242–44, 245n5

WAZA. *See* World Association of Zoos and Aquariums
White, Gilbert, 96

Why Zoos and Aquariums Matter study, 111
wild: absence of humans perceived as, 116, 117; ASDM survey on concept of, 116–20, 124n6; domestication as not opposite of, 145–47; idealized, 122; irony of close animal-human interactions as, 119, 121; natural v., 179; nuanced understandings of, 116–17; safety concerns, 118; vernacular usage of, 48. *See also* zoos, wilder; zoo-wild relationship
wild, American Indian understanding of: American Indian social and historical context, 81–85; bear-person and agency example of, 80; civilizing "wild" Indians, 84; as fearless, 80; as freedom, 88; irony of, 88; language and, 80–81; Navajo reflections and playful use of, 78–79; noble savage trope, colonialism and, 81–85; power dynamic in, 78; questions for further understanding, 89; relationality in, 77; settler view of noble side of, 88; Skidi-Pawnee and Tohono O'odham upbringing and, 79–80, 89n1; unfreedom, savage side of, 88; zoo displays of natives, "close to nature" and, 85–86
wild, the: acknowledging human influence on, 72–73; complex meanings of current life and, 60–61; constant change in, 129–30; coyotes in Seattle and, 60–61; diminished access and increased demand for, 97; duality and artificial separation from, 71, 72; enclosure of, 93–95, 101; graded view of, 12; hands-off position and policies on, 72; history of institutionalization of, 95–98; human impact and definition of, 11; individual animal stories and, 59; misconceptions about dogs and, 142–47; prompts, provocations and place for considering, 25–26; workshop discussion on, 107–8, 111; youth false sense of, 98–99; zoo staff time with, 103
Wilder Kingdom workshop, 58–59, 107–8, 111
wilderness, 7; civilization supremacy and, 87; Cronon (historian) on, 117; disappearing natives and, 86; human threat to values of, 23; mindsets for wildness and, 18–19; "no human impact" definition of, 16; *self-willed*, *untrammeled* diagnostics of, 18–19
Wilderness Act in 1964, 7, 12
Wilderness and the American Mind (Nash), 7
"Wilderness Letter" (Stegner), 1
wilder zoos. *See* zoos, wilder
Wild Kingdom, clichés and, 103

Wildlife Alliance, 231–32
Wildlife Conservation Society, 38
wildness (wild-ness): agency and, 37, 80, 88–89; animal behavior as, 116; animal demonstrations as, 117–18; architecture, academicization and, 193–94; change in understanding of, 14–15; Christian EuroAmerican settlers and, 82–83; human-inclusive definition of, 213–14; human participation in, 18; human threat to values of, 23; limitations in reactions to, 67–68; managed, 197–98; mindsets for wilderness and, 18; natural history and, 15–22; "peaceful," 20–21; philosophical underpinnings, 20; prompts and place for considering, 25–26; rapid evolutionary change impact for, 126; real and imagined, 14–24; responsibility and, 241–45; *self-willed* and *untrammeled* diagnostics of, 18–19; size and, 169–70; valuing, 65–68; zoo encounters with, 68–71. *See also* zoos, wilder
Wilson, E. O., 99
Wisconsin, cranes in, 35
wolves: becoming dogs, 142–43; tamed, 146; zoos exhibiting Canidae, subspecies of, 141, 151n9
wonder, grief and, 59
Woodland Park Zoo, 2, *10*, 163–64
Wordsworth, William, 60
World Association of Zoos and Aquariums (WAZA), 186–87, 193
World of Darkness, Bronx Zoo, 158

Yellowstone National Park, 197–98
Yosemite National Park, 198

zoo biology, 164
Zoo Biology (Conway), 198–200
zoo design, 1; ASDM, 26; based on animals, 164–66; connection goal and, 63–64, 68–71; connection through landscape emphasis in, 70; ecosystems and environments in, 68–69, 70; Hancocks on, 13–14; immersive and naturalistic, 9, 11, 163–64; minirevolutions in, 9, 11, 163–64; "museums without walls" vision for, 103; nocturnal exhibits dead end for, 160; "quasi wild," 11; sanitary approach to, 8; technology incorporation in, 70–71; visibility, temporality and, 165–66; Woodland Park Zoo, 2. *See also* architecture, zoo and aquarium
Zoologies (Deming), 50–51
zoos: accredited, 110; art activities in, 60; art in, 52; ASDM categorization as, 112, 114–15; captive populations adaptation to, 127; CEOs of, 200; cheetah companion dogs in, 139, *140*; as communities, 39; connection goal of, 63–64, 68; conservation

zoos (*continued*)
budget of, 102–3; conservation history absence of, 7–8, 28n16; conservation involvement of, 231; Conway critique of modern, 198–200; critics of, 2; culture building in, 193–94; death of iconic species in early, 101; debates about, 3, 214–15; development of, 95–98; dichotomy exposed by, 2; dilemma faced by, 188; dogs absence in, 140–41, 147; drive-through, 116; as educational environments, 186, 193–94; empathy/conservation and, 199; encounters with wild at, 68–71; ex situ-in situ integration necessity, 236–39; feeding time in, 160–63; fieldwork attitude in, 203; first containing large wild animals, 44n7; first modern, 36, 44n5; first use of term, 35; future, 103–4; hybrid institutions and, 43n4; importance of aquariums and, 111; institutional crises, 93–95; messaging of, 232–36; moralizing about, 12–13; natives/colonial subjects on display in, 85–86; "natural" and "wild," 11; nature/culture dichotomy of, 87; nature understanding encouraged by, 14; need for new stories, 103–4, 213–14, 223–25; nocturnal exhibits in, 156–60, *157*, *159*; pandemic impact on, 188; portal to wild image of, 242; public notion of characteristics, 112, *114*, 114–15, 123n4; purpose of, 49–50, 63–64; reforms to cages/enclosures in, 219–20; regional conservation example, 209–10; trains and, 155–56; unusual dogs exhibited in, 141, 151n9; wolf subspecies in (Canidae), 141, 151n9. *See also* architecture, zoo and aquarium; cages and enclosures; zoo design; *specific zoos*

zoos, wilder, 2, 5, 9; agency, relationships in creating, 88–89; architectural tasks in context of, 183–84; art project on, 58–59; ASDM and, 118–20, 122; changing meaning of "wilder," 59; compassion question on, 122; Conway conception of, 198–200; enclosure size and, 118–19; endangered species and, 242; financial restraints, 201–2; parks and, 219–26; questions to consider on, 24; regional conservation programs for, 202–4, 208–9; relationality and, 241; relational meaning of "wilder," 59; revolution needed for, 163–66; in situ conservation support requirement, 241–42; zoo purpose and, 49–50

zoos and parks: connections and separations in, 216–19; new stories and connections for, 223–25; shared qualities of, 216; wilder, 219–26
Zootopia, in Denmark, 9
zoo-wild relationship, 1; environmental ethics and, 4, 7; ICF as case study on, 33–34; pragmatic preservation outlook on, 4–14
Zoo Zurich, development plan for, 192, *192*
Zurich Zoo: elephant enclosure in, 190; Madagascar Rainforest exhibit, 164

Printed and bound by CPI Group (UK) Ltd, Croydon, CR0 4YY

17/06/2024

14516439-0002